2

EUROPEAN POLITICAL, ECONOMIC, AND SECURITY ISSUES

EUROPEAN ENERGY SECURITY OPTIONS AND CHALLENGES

EUROPEAN POLITICAL, ECONOMIC, AND SECURITY ISSUES

Additional books in this series can be found on Nova's website under the Series tab.

Additional E-books in this series can be found on Nova's website under the e-books tab.

ENERGY POLICIES, POLITICS AND PRICES

Additional books in this series can be found on Nova's website under the Series tab.

Additional E-books in this series can be found on Nova's website under the e-books tab.

EUROPEAN POLITICAL, ECONOMIC, AND SECURITY ISSUES

EUROPEAN ENERGY SECURITY OPTIONS AND CHALLENGES

RADCLIFF E. PHILLIPS
AND
JAMISON B. COOK
EDITORS

New York

For permission to use material from this book please contact us:
Telephone 631-231-7269; Fax 631-231-8175
Web Site: http://www.novapublishers.com

LIBRARY OF CONGRESS CATALOGING-IN-PUBLICATION DATA

ISBN: 978-1-62081-814-5

Published by Nova Science Publishers, Inc. † New York

CONTENTS

PREFACE

Europe as a major energy consumer faces a number of challenges when addressing future energy needs. Among these challenges are a rapidly rising global demand and competition for energy resources from emerging economies such as China and India, persistent instability in energy producing regions such as the Middle East, a fragmented internal European energy market, and a growing need to shift fuels in order to address climate change policy. As a result, energy supply security has become a key concern for European nations and the European Union (EU). This book focuses on potential approaches that Europe might employ to diversify its sources of natural gas supply, and Russia's role, as well as identifying some of the issues hindering efforts to develop alternative suppliers of natural gas.

Chapter 1 - Europe as a major energy consumer faces a number of challenges when addressing future energy needs. Among these challenges are a rapidly rising global demand and competition for energy resources from emerging economies such as China and India, persistent instability in energy producing regions such as the Middle East, a fragmented internal European energy market, and a growing need to shift fuels in order to address climate change policy. As a result, energy supply security has become a key concern for European nations and the European Union (EU). A key element of the EU's energy supply strategy has been to shift to a greater use of natural gas. Europe as a whole is a major importer of natural gas. Russia is Europe's most important natural gas supplier, accounting for 34% of Europe's natural gas imports. Europe's natural gas consumption is projected to grow while its own domestic natural gas production continues to decline. If trends continue as projected, Europe's dependence on Russia as a supplier is likely to grow. And, while it could be in Europe's interest to explore alternative sources for its natural gas needs, it is uncertain whether Europe as a whole can, or is willing to, replace a significant level of imports of Russian natural gas. Some European countries that feel vulnerable to potential Russian energy supply manipulation may work harder to achieve diversification than others.

Chapter 2 - This is a statement of Honorable Gregory W. Meeks, United States House Representative.

Chapter 3 - This is a testimony of Ambassador Richard L. Morningstar, Special Envoy for Eurasian Energy.

Chapter 4 - This is a testimony of Ariel Cohen, Senior Research Fellow for the Heritage Foundation.

Chapter 5 - This is a testimony of Keith C. Smith, Center for Strategic and International Studies.

Chapter 6 - This is a testimony of Ambassador Ross Wilson, Atlantic Council of the United States.

Chapter 7 - The Russian oil and natural gas industries are key players in the global energy market, particularly in Europe and Eurasia. Another trend has been the concentration of these industries in the hands of the Russian government. This latter phenomenon has been accompanied by an authoritarian political system, in which former intelligence officers play key roles. Russian firms have tried to purchase a controlling stake in pipelines, ports, storage facilities, and other key energy assets of European countries. They need these assets to transport energy supplies to lucrative western European markets, as well as to secure greater control over the domestic markets of the countries of the region. In several cases where assets were sold to non-Russian firms, Russian firms cut off energy supplies to the facilities. Russia has also tried to build new pipelines to circumvent infrastructure that it does not control. Another objective Russia has pursued has been to eliminate the energy subsidies former Soviet republics have received since the fall of the Soviet Union, including by raising the price these countries pay for natural gas to world market prices.

In: European Energy Security Options ... ISBN: 978-1-62081-814-5
Editors: R. E. Phillips and J. B. Cook

Chapter 1

EUROPE'S ENERGY SECURITY: OPTIONS AND CHALLENGES TO NATURAL GAS SUPPLY DIVERSIFICATION*

Michael Ratner, Paul Belkin, Jim Nichol and Steven Woehrel

SUMMARY

Europe as a major energy consumer faces a number of challenges when addressing future energy needs. Among these challenges are a rapidly rising global demand and competition for energy resources from emerging economies such as China and India, persistent instability in energy producing regions such as the Middle East, a fragmented internal European energy market, and a growing need to shift fuels in order to address climate change policy. As a result, energy supply security has become a key concern for European nations and the European Union (EU).

A key element of the EU's energy supply strategy has been to shift to a greater use of natural gas. Europe as a whole is a major importer of natural gas. Russia is Europe's most important natural gas supplier, accounting for 34% of Europe's natural gas imports. Europe's natural gas consumption is projected to grow while its own domestic natural gas

* This is an edited, reformatted and augmented version of a Congressional Research Service publication, CRS Report for Congress R42405, prepared for Members and Committees of Congress, from www.crs.gov, dated March 13, 2012.

production continues to decline. If trends continue as projected, Europe's dependence on Russia as a supplier is likely to grow. And, while it could be in Europe's interest to explore alternative sources for its natural gas needs, it is uncertain whether Europe as a whole can, or is willing to, replace a significant level of imports of Russian natural gas. Some European countries that feel vulnerable to potential Russian energy supply manipulation may work harder to achieve diversification than others.

Russia has not been idle when it comes to protecting its share of the European natural gas market. Moscow, including the state-controlled company Gazprom, has attempted to defeat European-backed alternatives to pipelines it controls by proposing competing pipeline projects and attempting to co-opt European companies by offering them stakes in those and other projects. It has attempted to dissuade potential suppliers (especially those in Central Asia) from participating in the European-supported plans. Moscow has also raised environmental concerns in an effort to stymie other alternatives to its supplies, such as unconventional natural gas.

Successive US administrations and Congresses have viewed European energy security as a US national interest. Promoting diversification of Europe's natural gas supplies, especially in recent years through the development of a southern European corridor, as an alternative to Russian natural gas has been the mandate of the State Department's Special Envoy for Eurasian Energy. The George W. Bush Administration viewed the issue in geopolitical terms and sharply criticized Russia for using energy supplies as a political tool to influence other countries. The Obama Administration has also called for diversification, but has refrained from openly expressing concerns about Russia's regional energy policy, perhaps in order to avoid jeopardizing the "reset" of ties with Moscow. Additionally, a change in tenor from the Obama Administration towards the Nabucco pipeline project may indicate waning interest in the southern corridor strategy.

This report focuses on potential approaches that Europe might employ to diversify its sources of natural gas supply, and Russia's role, as well as identifying some of the issues hindering efforts to develop alternative suppliers of natural gas. The report assesses the potential suppliers of natural gas to Europe and the short- to medium-term hurdles needed to be overcome for those suppliers to be credible, long-term providers of natural gas to Europe. The report looks at North Africa, probably the most realistic supply alternative in the near-term, but notes that the region will have to resolve its current political and economic instability as well as the internal structural changes to the natural gas industry. Central Asia, which may have the greatest amounts of natural gas, would need to construct lengthy pipelines through multiple countries to move its natural gas to Europe.

Introduction: Change Is Afoot

The 27 member-state European Union (EU) has been a growing natural gas consumer and importer for decades. However, as Europe's natural gas production has declined in recent years, its dependence on imported natural gas has increased. This has left it more dependent as a whole on its primary supplier, Russia, which has shown some inclination to use its resources for political ends. Natural gas, unlike oil which is a global commodity, is a regional commodity with regional buyers and sellers exerting more influence.

Over the past decade, some European officials have become increasingly concerned about the potential for cutoffs or curtailments of Russian natural gas supplies to Europe. Most Russian natural gas exports to Europe flow through Ukraine and Belarus. Fragile and sometimes hostile relations between Kyiv, Minsk, and Moscow have in the past resulted in interruptions in the flow of natural gas to parts of Europe, as happened in 2006 and 2009. Some countries in Eastern Europe, which are in some cases almost exclusively reliant on Russian gas imports, have been particularly susceptible to these fluctuations.

In response to past supply cutoffs and the potential for future energy supply interruptions European leaders, sometimes with the support of the United States, have sought to take steps to increase their energy security by exploring new supply diversification options. One such response, though contrary to the US perspective of energy security through diversification, has been Germany's decision to support construction of the Nord Stream pipeline, which directly connects Russia and Germany, Russia's largest importer. Russia has also committed to building the South Stream pipeline across the Black Sea, connecting Russia, Bulgaria, and Hungary. While these pipeline projects bypasses transit states such as Ukraine and Belarus, they also bypass EU member states like Poland and Lithuania that are more critical of Russian policies and also present rivals to other pipelines being pursued by the EU.

The opening of Nord Stream and the proposal for South Stream highlights a challenge Europe faces with diversifying its natural gas supplies: Russia has demonstrated a willingness to go to great lengths to maintain its hold on European market share of natural gas. However, while some European countries, Germany included, maintain that projects such as Nord Stream enhance European security by providing alternate routes for Russian supplies, a number of EU member states, including Poland and Lithuania, opposed Nord Stream and have questioned Russia's reliability as an energy supplier. Critics tend to argue, for example, that projects like Nord Stream could give Moscow additional political and economic leverage in its dealings with countries that have been bypassed by the pipeline. Meanwhile, alternative supplies from other regions (e.g., North Africa and Central Asia) face several significant challenges.

A second EU response to concerns over Europe's reliance on Russian natural gas supplies is what has become known as the Southern strategy. The so-called Southern Gas Corridor's flagship initiative is the proposed Nabucco natural gas pipeline. The pipeline is intended to transport up to 1.1 trillion cubic feet of Caspian (and perhaps Middle Eastern) natural gas per year through Turkey into Bulgaria and on to Austria. The Nabucco project has, however, been beset by lengthy delays and ongoing questions about its viability. The six-company Nabucco consortium hopes to begin construction in 2013, but many observers question this timeframe, citing ongoing concerns about capacity and funding.A third European energy initiative involves Europe's own fragmented internal energy market. In early February 2011, European heads of state pledged to: complete the integration and liberalization of the internal European energy market by 2014; ensure all European member states are connected to a Europe-wide energy supply grid by 2015; boost energy efficiency throughout Europe; and better coordinate external energy policies. European leaders hope that further market liberalization and interconnection of electric grids and pipelines will, among other things, allow member states to share and trade energy more flexibly than at present, mitigating the impact of supply interruptions and overdependence on a single supplier. The European Commission estimates that over €1 trillion (about $1.4 trillion) of infrastructure and other investment will be necessary to realize the EU goals.

Despite its growing dependence on Russian natural gas, Europe is well positioned geographically to benefit from recent changes in global natural gas development. Since the advent of shale gas in the United States, the world appears to be potentially awash in natural gas. A 2011 study commissioned by the US Energy Information Administration (EIA) showed that technically recoverable shale gas resources worldwide may exceed current global natural gas reserves.[1] Other key developments and possible alternatives to Russian natural gas are outlined below:

- Taken as a whole, North Africa could pose a credible alternative to Russian natural gas supplies. The change of regimes in Libya, in particular, and in Egypt as a result of the wave of regional unrest known as the "Arab Spring," poses a potential opportunity to increase natural gas production and exports from these countries. Both Libya and Egypt have large natural gas reserves, but production and exports have been hampered by domestic policies. Algeria, the largest exporter of natural gas in North Africa and the third largest supplier to Europe behind Russia and Norway, may also hold large volumes of shale gas yet to be developed in addition to their substantial conventional reserves.

- Central Asia may hold the greatest potential for new natural gas supplies for Europe, but currently those supplies would have to transit Russia to arrive in the European market. The delays in developing a southern corridor natural gas pipeline route to Europe have forced Central Asian countries to look east instead of west to bypass Russia and open new markets.[2]
- Liquefied natural gas (LNG) imports pose an additional alternative to Russian natural gas. In 2010, LNG comprised almost 20% of the EU's natural gas imports and over 15% of its consumption. The EU has LNG import capacity to meet its peak winter demand for natural gas, but during most of the year the facilities are underutilized. Nevertheless, some countries are considering building additional LNG import terminals to diversify their sources of natural gas. In addition to LNG import terminals, the EU could benefit from increased natural gas storage facilities in order to manage their import capacity during non-peak periods, as well as more pipeline interconnections to move natural gas where it is needed. EU officials have identified both improvements as priorities and they are being pursued, but not without some difficulty.

A third European energy initiative involves Europe's own fragmented internal energy market. In early February 2011, European heads of state pledged to: complete the integration and liberalization of the internal European energy market by 2014; ensure all European member states are connected to a Europe-wide energy supply grid by 2015; boost energy efficiency throughout Europe; and better coordinate external energy policies. European leaders hope that further market liberalization and interconnection of electric grids and pipelines will, among other things, allow member states to share and trade energy more flexibly than at present, mitigating the impact of supply interruptions and overdependence on a single supplier. The European Commission estimates that over €1 trillion (about $1.4 trillion) of infrastructure and other investment will be necessary to realize the EU goals.

Despite its growing dependence on Russian natural gas, Europe is well positioned geographically to benefit from recent changes in global natural gas development. Since the advent of shale gas in the United States, the world appears to be potentially awash in natural gas. A 2011 study commissioned by the US Energy Information Administration (EIA) showed that technically recoverable shale gas resources worldwide may exceed current global natural gas reserves.[1] Other key developments and possible alternatives to Russian natural gas are outlined below:

- Taken as a whole, North Africa could pose a credible alternative to Russian natural gas supplies. The change of regimes in Libya, in particular, and in Egypt as a result of the wave of regional unrest known as the "Arab Spring," poses a potential opportunity to increase natural gas production and exports from these countries. Both Libya and Egypt have large natural gas reserves, but production and exports have been hampered by domestic policies. Algeria, the largest exporter of natural gas in North Africa and the third largest supplier to Europe behind Russia and Norway, may also hold large volumes of shale gas yet to be developed in addition to their substantial conventional reserves.
- Central Asia may hold the greatest potential for new natural gas supplies for Europe, but currently those supplies would have to transit Russia to arrive in the European market. The delays in developing a southern corridor natural gas pipeline route to Europe have forced Central Asian countries to look east instead of west to bypass Russia and open new markets.[2]
- Liquefied natural gas (LNG) imports pose an additional alternative to Russian natural gas. In 2010, LNG comprised almost 20% of the EU's natural gas imports and over 15% of its consumption. The EU has LNG import capacity to meet its peak winter demand for natural gas, but during most of the year the facilities are underutilized. Nevertheless, some countries are considering building additional LNG import terminals to diversify their sources of natural gas. In addition to LNG import terminals, the EU could benefit from increased natural gas storage facilities in order to manage their import capacity during non-peak periods, as well as more pipeline interconnections to move natural gas where it is needed. EU officials have identified both improvements as priorities and they are being pursued, but not without some difficulty.

- The prospect of significant US LNG exports may pose an opportunity for the United States to play a bigger role in European energy security and global natural gas markets.[3] Most of the proposed US LNG export projects are located on the Gulf coast or east coast of the United States, making shipments, at least initially, more likely to go to Europe than Asia. Additionally, the US natural gas market is one of the only markets in the world where natural gas is not priced against oil, giving it a cost advantage in most of Europe. Should future US LNG contracts not include an oil-indexed formula, pressure would be added for other countries, including Russia, to follow suit. Russian companies, including state-controlled natural gas giant Gazprom, have adamantly defended oil-indexed natural gas prices.

CONTEXT, BACKGROUND AND DIFFERENT POINTS OF VIEWS

The US Perspective

The primary focus of US efforts has been on establishing a southern corridor route for Central Asian and Middle Eastern natural gas supplies to be shipped by pipeline to Europe. Other efforts have been focused on EU market reforms, which are beyond the scope of this paper.

The George W. Bush Administration viewed the issue in geopolitical terms and sharply criticized Russia for using energy supplies as a means to gain political influence over other countries.[4] The Obama Administration has also called for diversification, but has refrained from openly expressing concerns about Russia's energy policy in the region, perhaps in order to avoid jeopardizing the "reset" of ties with Moscow. Additionally, a change in tenor from the Obama Administration towards the Nabucco pipeline project may indicate waning interest in the southern corridor strategy.

Regarding Central Asian supplies and European energy security, in testimony to Congress in June 2011, Richard Morningstar, the State Department's Special Envoy for Eurasian Energy, said that US policy encourages the development of new Eurasian oil and natural gas resources to increase the diversity of world energy supplies. A second US goal is to increase European energy security, so that some countries in Europe that largely rely on a single supplier (presumably Russia) may in the future have diverse suppliers. A third goal is assisting Caspian regional states to develop new routes to market, so that they can obtain more competitive prices and become more prosperous. In order to achieve these goals, the Administration supports the development of the Southern Corridor of Caspian (and perhaps Iraq) natural gas export routes transiting Turkey to Europe.[5] Of the three main vying pipeline consortia—the Nabucco, the Interconnector-TurkeyGreece-Italy (ITGI), and the Trans-Adriatic Pipeline (TAP) groups—the Obama Administration has said it will support the project "that brings the most gas, soonest and most reliably, to those parts of Europe that need it most."[6] However, given the historically close relationship between Russia and Italy on energy issues, including natural gas (especially under the government of former prime minister Silvio Berlusconi), supporting projects that terminate in Italy could also raise concerns about Russian influence. Such concerns could be mitigated depending on the policy measures taken by future Italian governments.

At the same time, Morningstar has rejected views that Russia and the United States are competing for influence over Caspian energy supplies, stating that the Administration has formed a Working Group on Energy under the US-Russia Bilateral Presidential Commission.[7] At a conference in Azerbaijan in November 2011, Morningstar stressed that while the United States still supports Nabucco, "if a smaller pipeline is chosen to ship [Azerbaijani] gas to Europe, we believe it should provide gas to vulnerable countries in Europe and there should be concrete, written guarantees that the pipeline will be expanded as more gas becomes available," ostensibly so that Europe is less dependent on Russia as a source.[8]

The Arab Spring, which has brought regime change to two large natural gas producers, Libya and Egypt, offers potentially expanded sources of natural gas to Europe, but it will depend upon the policies of the evolving governments. North Africa already has significant natural gas infrastructure—LNG export terminals and pipelines—connecting it to Europe. However, it is too early to determine how the changes in North Africa and the Middle East (MENA) will affect natural gas production and exports. The US government, along with the EU, has indicated its desire to expand trade and investment with the MENA region, which could help foster economic growth and provide support for successful democratic transitions. For example, in a speech delivered at the State Department on May 19, 2011, President Obama outlined a new plan for US engagement with MENA.[9] A key part of this plan is launching a "Trade and Investment Partnership Initiative" with the MENA countries. Some Members of Congress have also expressed interest in deeper trade and investment ties with Arab Spring countries.[10]

Although US trade and investment with the MENA region overall is relatively limited at present, this region may present growing commercial opportunities for US businesses in areas such as energy, transportation, and infrastructure.[11]

The 112th Congress has also expressed concern about European energy security. Section 1233 of the FY2012 National Defense Authorization Act requires the Secretary of Defense to submit to "the appropriate committees of Congress a detailed report on efforts by the Department of Defense, including within NATO, to address the energy security of the NATO alliance."

European Natural Gas Consumption and the EU's Evolving Energy Policy

Collectively, EU member states are the world's largest energy importer, importing about 55% of their energy supply—approximately 84% of their oil and 64% of their natural gas.[12] EU member states increasingly rely on natural gas, particularly to reach ambitious targets to reduce carbon dioxide and greenhouse gas emissions. Natural gas comprised over 25% of the EU's primary energy consumption in 2010, a number that is expected to grow to almost 30% by 2030.[13] Oil made up almost 40%, coal about 16%, and nuclear 12% of the EU primary energy supply.

The European Commission forecasts that the EU will import over 80% of its natural gas needs by 2030. Analysts note that recent policy decisions, such as a 2011 German announcement that it would phase out use of its nuclear power plants by 2020 and a French decision to prohibit shale gas development, could mean a more rapid rise in Europe's dependence on natural gas imports. Other EU countries have made similar announcements, but are much smaller energy consumers.

Russia has long been, and is expected to continue to be, the key supplier of natural gas to Europe. In 2010, Russia accounted for 34% of European natural gas imports, followed by Norway and Algeria (see *Figure 1*). Russian and European companies have developed an extensive network of infrastructure to transport Russian natural gas long distances to European markets.

Observers expect natural gas to play a significant role in Europe-Russia relations for decades to come.

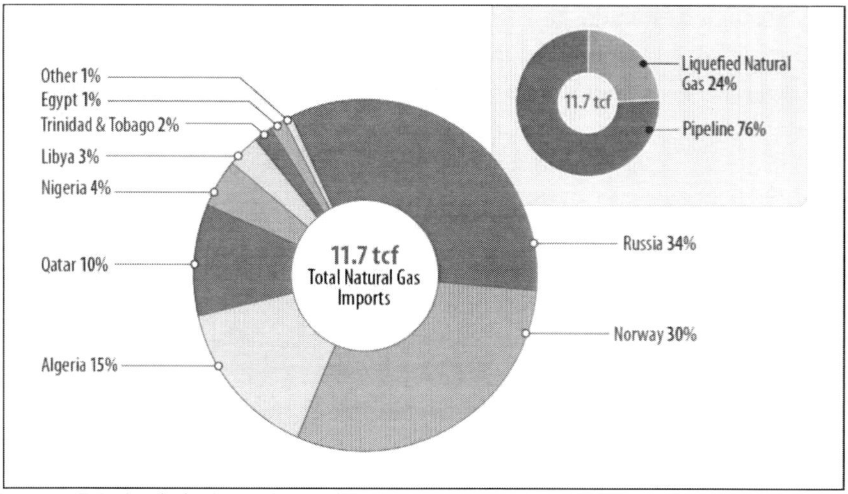

Source: BP Statistical Review of World Energy 2011, http://www.bp.com/section genericarticle800.do? categoryId=9037130&contentId=7068669.

Notes: The United States re-exported a minimal amount of LNG to Europe in 2010 and is included in Other. Units are trillion cubic feet (tcf).

Figure 1. 2010 EU Natural Gas Imports.

Different EU member states use natural gas to different degrees and import levels and sources vary by country (see *Table 1*). Some large natural gas consumers, such as Spain, do not import any natural gas from Russia. Germany, the second biggest natural gas consumer and Russia's largest market, relied on Russia for almost 40% of its imports in 2010. The opening of the Nord Stream pipeline in late 2011 and Germany's planned closure of its nuclear power plants highlights Germany's potentially greater reliance on Russia.

In a reflection of these national differences, the EU has traditionally exerted little if any influence over the energy policies of individual member states. However, in the face of rising concern about Europe's reliance on Russian energy and growing public pressure to address global climate change, EU member states have begun to increase cooperation toward an "Energy Policy for Europe." As stated earlier, European heads of state have committed to completing the integration and liberalization of the internal European energy market by 2014; promoting the interconnection of electric grids and natural gas pipelines; boosting energy efficiency; and better coordinating external energy policies. European leaders anticipate that these initiatives will allow member states to share and trade energy more flexibly than at present, mitigating the impact of potential supply interruptions and overdependence on a single supplier.

Even as EU leaders promote ideas on a common energy strategy, many question how far individual member states will agree to push Russia (and Gazprom) to adopt the EU's principles of competition and open its energy sector to outside investment. Some believe that without such Russian concessions, Europe will ultimately find its energy security largely under Russian control. Indeed, several member states have pursued bilateral energy deals with Russia that will increase their dependence on Russia for years to come. Both Germany and Italy, the largest importers of Russian natural gas, have negotiated long-term deals with Russia to lock in future natural gas supplies. For Germany and several others, Russia's role as a dominant energy supplier increases the importance of fostering good relations with Moscow. Further, bilateral deals with Russia are not limited to the major energy consumers. Bulgaria, Romania, Hungary, Greece, and others have entered into long-term energy agreements with Gazprom over the past several years.

These examples of individual member states dealing with Russia bilaterally have in the past drawn harsh criticism from other EU member states, such as the Baltic states and Poland, who have had strained relations with Russia for some time over other issues as well. Governments in these countries have warned their European colleagues not to make energy deals that could give Russia increased political influence over European decision-making. Many of these nations believe that Europe's dependence on Russian energy is likely to last no matter how successful Europe may be in identifying energy supply alternatives. But they also feel Europe does not gain real security by becoming more dependent on Russia. In fact, the growing presence of Gazprom throughout the European energy market (for instance through its ownership of distribution and storage infrastructure) has led many to worry about the EU's ability to develop an energy policy insulated from Gazprom's influence.[14]

The EU's 2011 decision to further liberalize Europe's energy market could signal the beginning of a more unified approach toward Russia. Moscow has strongly criticized the decision, which, among other things, would require energy companies that own pipelines to sell them, or manage them separately. Under the policy, Gazprom, which plays a key role in exporting natural gas to Europe, could be forced to sell its significant stakes in European distribution networks. In December 2011, Gazprom announced that its South Stream natural gas pipeline would end (discussed in more detail in "Russia's Role") in Italy rather than in Austria, as was previously planned. Company sources reportedly stated that the change was in reaction to an EU decision to block a Gazprom bid to purchase a 50% stake in the Central European Gas Hub (CEGH) in Austria.[15] EU member states have committed to fully implementing the liberalization directive by 2014. However, as of late 2011, at least 18 countries had yet to transform the directive into national law.[16]

EU and US concerns about overdependence on Russian natural gas supplies have also led to the proposed Nabucco gas pipeline. Nabucco would run from eastern Turkey to Bulgaria, Romania, Hungary, and Austria and on to other European countries. Natural gas would be supplied primarily by Azerbaijan via pipelines transiting through Georgia. The Nabucco consortium had hoped to begin work on the pipeline in 2010, but the project has faced repeated delays. A key factor is uncertainty about Azerbaijan's production capacity and whether Azerbaijani supplies could alone justify building the pipeline. The participation of Turkmenistan and Iraq could be key to Nabucco's success, but it could be years before these countries are ready to ship gas through the proposed pipeline. Some Nabucco advocates in Europe have also raised the possibility of using the pipeline to transport Iranian gas, a remote alternative given current sanctions. The lingering effects of the global financial crisis could also make it difficult for the private sector to finance Nabucco and other projected pipelines. Indeed, reports in 2011 suggested that the total cost of the project could be almost double the €7.9 billion estimated by the consortium of European energy companies that have committed to build the Nabucco pipeline.[17]

Russia's Role[18]

The Russian natural gas industry is one of the most important players in the global energy market. In 2010, Russia had the largest natural gas reserves in the world, nearly 24% of the world's total, was the leading exporter of natural gas, and placed second in production and consumption behind the United States. Russia was also a founding member, and currently holds the top position, in the Gas Exporting Countries Forum (GECF).

As noted, Russia is currently the dominant supplier of natural gas to Europe, accounting for about one-quarter of the EU's natural gas supplies.[19] (See *Figure 2.*) This dependency does not go only in one direction, however. Europe is also the most important market for Russian natural gas exports, a calculation the Russians must take into account when developing its political relations with Europe.

Table 1. EU Natural Gas Data, 2010 Units equal billion cubic feet per year (bcf)

	Natural Gas Consumption	*Natural Gas Production*	*Natural Gas Imports*
Austria	297	58	239
Belgium	738	0	867[a]
Bulgaria	77	<1	76
Cyprus	0	0	0
Czech Republic	415	7	408
Denmark	172	292	5
Estonia	13	0	13
Finland	159	0	159
France	1,699	26	1,727
Germany	3,206	449	3,278[a]
Greece	137	0	137
Hungary	352	88	264
Ireland	198	11	187
Italy	2,949	293	2,661
Latvia	23	0	23
Lithuania	93	0	93
Luxembourg	48	0	48
Malta	0	0	0
Netherlands	1,750	3,034	599[a]
Poland	508	150	358
Portugal	175	0	175
Romania	446	370	76
Slovakia	197	4	193
Slovenia	31	<1	31
Spain	1,248	4	1,285[a]
Sweden	58	0	58
United Kingdom	3,329	1,988	1,894[a]
TOTAL	*18,317*	*6,773*	*14,855*

Sources: BP Statistical Review of World Energy 2011, Cedigaz, Eurogas, and the European Commission.

Notes: Imports plus internal production does not equal consumption because some countries export imported natural gas or their own production within the region.

[a]. Some EU countries import more natural gas than they require in order to re-export the natural gas to other countries.

The bulk of Gazprom's natural gas exports go to Europe and Eurasia. Of the 7.1 trillion cubic feet (tcf) of natural gas exported by Gazprom in 2010, almost 55% went to the EU. Of the rest, over 28% went to the Commonwealth of Independent States (CIS), many of which have been unreliable in paying what they owe and/or receive natural gas at subsidized prices.[20] The rest went to Turkey, which is seeking EU membership, and other non-EU countries in Europe, and to Asia.[21]

The Gas Exporting Countries Forum

The Gas Exporting Countries Forum (GECF), also known as Gas-OPEC, is composed of some of the world's leading natural gas producers and exporters. It is not a cartel in the same sense as OPEC, in that it does not control marginal production in an effort to influence prices. There are structural differences in global natural gas and global oil that make this type of control difficult. Nevertheless, the GECF provides a venue for its members to discuss topics of interest such as production projects, exports, etc. Its members—which include Algeria, Bolivia, Egypt, Equatorial Guinea, Iran, Libya, Nigeria, Qatar, Russia, Trinidad and Tobago, and Venezuela—control 36% of world production and 46% of global trade. Kazakhstan, the Netherlands, and Norway have observer status at the GECF. Major natural gas producers that are not affiliated with the GECF include Australia, Azerbaijan, Canada, Indonesia, Malaysia, Oman, Turkmenistan, the United States (the world's leading natural gas producer), and the United Arab Emirates, which collectively control 33% of world production and 28% of global trade.

The revenues generated by this trade are vital to the ruling Russian elite. At present, all Russian natural gas exports are controlled by Gazprom. As a state-controlled firm, Gazprom has the closest possible links with top Russian leaders (Russia's outgoing president Dimitri Mevedev served as president of Gazprom).

The personal and political fortunes of Russia's leaders are closely tied to Gazprom. Russian government revenues (in 2010, 46% of total Russian government revenue came from oil and natural gas taxes) and Russia's economic revival in the Putin/Medvedev era have been heavily dependent on the massive wealth generated by energy exports to Europe. Gazprom offers natural gas to the Russian domestic market at subsidized prices, which also bolsters the ruling elite politically. Government proposals to decrease subsidies have not come to fruition.

EU Energy Consumption of Russian Natural Gas (%)

	Primary Energy	Natural Gas
Austria	14.2%	62.5%
Belgium	0.0%	0.0%
Bulgaria	10.8%	99.5%
Cyprus	0.0%	0.0%
Czech Republic	18.4%	71.9%
Denmark	0.0%	0.0%
Estonia	6.0%	100.0%
Finland	13.9%	100.0%
France	2.9%	16.7%
Germany	9.7%	37.9%
Greece	5.7%	52.8%
Hungary	24.9%	65.0%
Ireland	0.0%	0.0%
Italy	7.4%	17.0%
Latvia	13.2%	100.0%
Lithuania	38.8%	100.0%
Luxembourg	0.0%	0.0%
Malta	0.0%	0.0%
Netherlands	3.6%	8.1%
Poland	8.5%	63.1%
Portugal	0.0%	0.0%
Romania	5.6%	17.0%
Slovakia	30.4%	98.2%
Slovenia	6.3%	56.2%
Spain	0.0%	0.0%
Sweden	0.0%	0.0%
United Kingdom	0.0%	0.0%
European Union	**5.7%**	**26.3%**

Source: Gas data from BP Statistical Review of Energy 2011, Cedigaz, Eurogas, European Commission; boundary data from ESRI, 2005. Graphic created by CRS.
Borders are not necessarily authoritative.

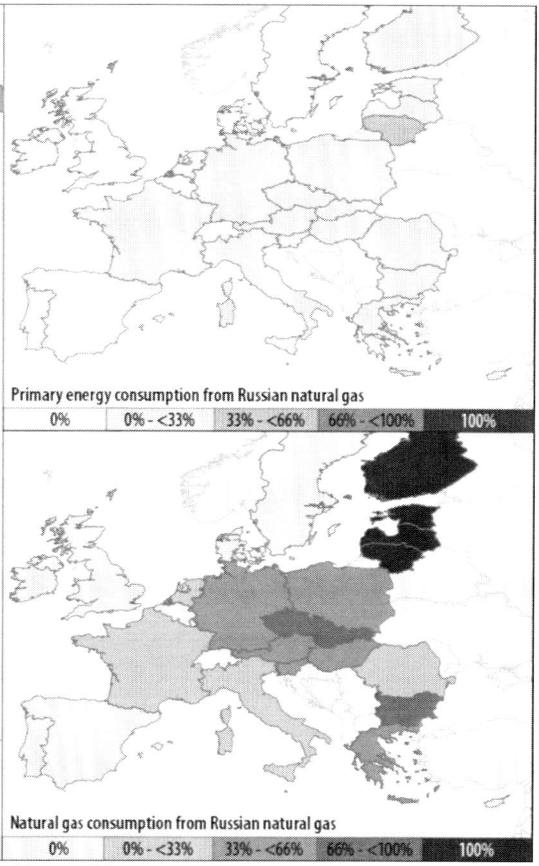

Source: CRS Graphics compiled this graphic.

Notes: For primary energy, which is the base source of energy used to produce electricity and perform other work, Russian natural gas does not comprise greater than 39% for any EU country.

Figure 2. EU Dependence on Russian Natural Gas.

In addition to their financial benefits, Russian natural gas exports to Europe and Eurasia may have important psychological benefits for the Russian elite. They may be viewed as demonstrating the resurgence of Russian power after the collapse of the Soviet Union over 20 years ago. Russia's "National Security Strategy to 2020," released in May 2009, stated that "the resource potential of Russia" is one of the factors that has "expanded the possibilities of the Russian Federation to strengthen its influence in the world arena."[22]

In the long term, Russia hopes to reduce dependency on Europe by diversifying its customer base as well. By 2030, the Russian government plans to increase gas exports to Asian countries such as China, South Korea, and Japan until they make up 19%-20% of the total. However, Russia has a considerable way to go to meet this objective. In 2010, gas exports to Asia made up about 7% of total Russian gas exports, all in the form of LNG. Russia opened its first LNG export facility in 2009 on its east coast. Long-standing Russian hopes of providing large amounts of natural gas to China by pipeline have been stymied by the fact that China has been unwilling to pay the price Europe pays for Russian natural gas.[23]

Given this situation, most experts believe that, barring the failure of Russia to increase its own energy exploration and development, Russia will continue to remain Europe's primary energy supplier, including natural gas supplies, for many years and possibly decades. And, Europe will remain the primary market for Russian energy exports. Therefore, the main goal of state-run Russian energy companies, such as Gazprom, has been to try to solidify their dominance of Europe's energy sector by pursuing long-term bilateral supply contracts with some European countries such as Germany, Italy, and Bulgaria, and by seeking to buy stakes in European energy distribution networks and storage facilities. Russia has also used the allure of its vast resources to co-opt European companies that dominate Europe's energy sector.

Gas Crises of the 2000s and Russia and Europe's Search for Alternatives

Although widely believed by industry and in some political circles, evidence that Russia has been able to exploit its energy strength to manipulate the policy of EU and other European countries is ambiguous. Some experts, particularly those in central Europe, claim that Russia is able to use its dominant role in the energy sectors of their countries to exert influence over certain businessmen and politicians. Others, mainly in western Europe, claim that the fact that Europe remains Russia's largest energy market, and thus its biggest source of foreign income, has led Russia to exercise more caution in dealing with EU countries. Key customers of Gazprom have been able to extract better contract terms in recent years that link part of the price of natural gas to spot natural gas prices instead of solely oil.

Russian leaders have repeatedly said that they view the former Soviet countries as lying within Russia's "sphere of privileged interests." Some have pointed out that Russia has openly used energy to affect domestic and international policies in Belarus and Ukraine. In perhaps the most striking example, Russia and Ukraine agreed to extend the stay of the Russian Black Sea Fleet in Crimea until 2042, from the original withdrawal date of 2017. In exchange, Russia pledged to provide Ukraine with a discount of two-thirds from Russia standard oil-linked contract price for natural gas supplies for 10 years. However, rising global oil prices (which have risen faster than spot natural gas prices), to which Russian contract prices are linked, have negated much of the savings Kyiv counted on, perhaps providing Moscow with additional leverage over Ukraine.[24]

In contrast, Russia may view countries such as Germany and France as key players on the world stage like itself, and therefore entitled to more respect. Smaller, former Soviet-controlled countries such as the Baltic and central European states may fall between these categories, in the view of Russian leaders.

In the mid- and late 2000s, many European countries suffered several unexpected energy cutoffs due to confrontations between Russia and the key pipeline transit states of Ukraine and Belarus over natural gas supply and transit issues. In 2009, Gazprom halted all natural gas supplies transiting Ukraine for nearly three weeks after the two sides failed to reach agreement on several issues, including a debt allegedly owed by Ukraine to Gazprom and the price that Ukraine would pay for natural gas supplies. Prior to the opening of Nord Stream, about 80% of Europe's natural gas imports from Russia transited Ukrainian pipelines. A similar Russian-Ukrainian dispute had led to a natural gas cutoff to Europe at the beginning of 2006. In 2010 and 2011, disputes between Russia and Belarus over a variety of issues, including energy prices, debts owed by Belarus, and transit fees paid by Russia for the use of Belarusian pipelines, led to temporary reductions of oil and natural gas supplies to Belarus and neighboring countries.

Russia and some western European countries responded to these incidents by planning new pipeline projects to bypass what they viewed as problematic transit states. One new natural gas pipeline is the aforementioned Nord Stream, which transports natural gas from Russia to Germany via a pipeline under the Baltic Sea. It has a planned capacity of almost 2 trillion cubic feet (tcf) per year, as compared to the Ukrainian pipeline system's 4.0-4.5 tcf per year. The first supplies from the pipeline were delivered in late November 2011, as stated earlier. Gazprom has proposed expanding Nord Stream's capacity still further, but Germany has rejected the idea so far.

Another pipeline project favored by Moscow is South Stream. It would run under the Black Sea to Bulgaria and then onto other European countries. Russia hopes to start construction of South Stream by the end of 2012, and begin deliveries in late 2015. South Stream has a planned capacity of 2.2 tcf per year and is considered a main competitor to the southern corridor projects, such as Nabucco and TAP (see "Central Asia Transit Constraints: Many

Options but No Alternatives So Far" below for more on the Southern Corridor projects).

While building pipelines that circumvent Ukraine, Russia continues its long-standing efforts to gain control of Ukraine's pipeline system. In fact, Russia is using Ukraine's fear of the potential impact of Nord Stream and South Stream on transit volumes and thus associated revenues through Ukraine's pipeline system to try to secure control of those pipelines cheaply. Gazprom officials have strongly encouraged Ukrainian leaders that they should sell control of Ukraine's pipelines to it while they can get a good price.[25] Otherwise, they say, Gazprom may find it more profitable to build and use South Stream rather than modernize Ukraine's aging system. Ukraine has offered Russia partial ownership of the Ukrainian pipeline system in exchange for a share in natural gas fields in Russia and guaranteed transit volumes through Ukraine's pipelines. So far Russia has not accepted Kyiv's terms. Russia has also rejected Ukraine's demands to renegotiate the current gas supply contract in order to cut the price Kyiv pays for gas, perhaps hoping that Ukraine's seeming desperation to secure lower gas prices could still induce it to give Gazprom de facto control over its pipelines.

Russia has had more success in gaining control of Belarus's gas infrastructure. In December 2011, Gazprom completed a deal to buy the 50% of Beltransgaz (Belarus's natural gas pipeline transport company) that it did not own, in exchange for reduced gas prices. The Yamal-Europe gas pipeline, which runs through Belarus and Poland, currently carries about 20% of Russian gas exports to Europe.

Some Russian actions may be aimed at frustrating European efforts at diversification.

These include trying to sign long-term contracts with Azerbaijan and Central Asian states to lock up supplies sought by the Europeans; lodging legal objections to the proposed Trans-Caspian Pipeline between Azerbaijan and Turkmenistan, which would be a key link in providing Caspian gas to Europe; attempting to coordinate natural gas export policies with other leading producers such as Qatar and Iran, perhaps with hopes of eventually creating a "gas OPEC" of the GECF; and the South Stream project itself, which may be as much an effort to thwart Nabucco as a viable pipeline project in its own right.

Central Asia Transit Constraints: Many Options but No Alternatives So Far

Establishing a non-Russian and non-Iranian natural gas pipeline system to transport Central Asian natural gas to Europe is a stated priority for the EU supported by the United States. Despite this, all the proposed projects face challenges from both cost and supply perspectives that raise questions about their viability (see *Table 2*). The main proposed pipeline, Nabucco, has had difficulty securing natural gas supplies and more recently cost questions have arisen.

Table 2. Prospective Non-Russian Southern Corridor Pipelines (Units = billion cubic feet per year (bcf))

Name	*Anticipated Capacity*	*Anticipated In-Service Date*	*Partners*
Interconnector Turkey, Greece, Italy (ITGI)	350	2015	BOTAS (Turkey), DEPA (Greece), Edison (Italy)
Trans Adriatic Pipeline (TAP)	350	2017	EGL (Switzerland), E.ON Ruhrgas (Germany), Statoil (Norway)
Trans-Anatolian Gas Pipeline (TANAP)	570	2017	BOTAS (Turkey), SOCAR (Azerbaijan), TPAO (Turkey)
South Caucasus Pipeline Expansion (SCP)[a]	565	2017	BP (United Kingdom), Lukoil (Russia), Naftiran (Iran), SOCAR (Azerbaijan), Statoil (Norway), Total (France), TPAO (Turkey)
South East Europe Pipeline (SEEP)	350	2017	BP (United Kingdom)
Nabucco Gas Pipeline	1,100	2019	BEH (Bulgaria), BOTAS (Turkey), MOL (Hungary), OMV (Austria), RWE (Germany), Transgaz (Romania)

Source: Company websites and various articles.

Notes: Not all of these projects will be built as many compete with each other for natural gas supplies. The South Stream pipeline project, Russia's response to developing the Southern Corridor for Caspian natural gas, is a 2,200 bcf per year pipeline sponsored by EDF (France), ENI (Italy), Gazprom (Russia), and Wintershall (Germany) to bring Russian natural gas to Europe. South Stream is also designed to bypass troubled transit states like Ukraine and Belarus.

[a]. The South Caucasus Pipeline (SCP) began operations in 2006 and currently has a capacity of 250 bcf of natural gas.

In mid-November 2007, Greek Prime Minister Kostas Karamanlis and Turkish Prime Minister Recep Tayyip Erdogan inaugurated a natural gas pipeline connecting the two countries. Since some Azerbaijani natural gas reaches Greece, the pipeline represents the first natural gas supplies from the Caspian region to the EU. If a pipeline extension is completed to Italy, this Interconnector Turkey-Greece-Italy (ITGI) natural gas pipeline could permit Azerbaijan to supply natural gas to two and perhaps more EU members, providing a source of supply besides Russia. The austerity measures undertaken by Greece may limit or prohibit the participation of DEPA, its state-owned natural gas company, in the ITGI pipeline.

The Nabucco pipeline has faced numerous delays, some of them attributable to Russia's counterproposals to build pipelines that it asserts would reduce the efficacy of the Nabucco pipeline and to questions about supplies for the pipeline. In early September 2010, the European Investment Bank, the European Bank for Reconstruction and Development, and the World Bank announced a commitment—pending environmental and social feasibility studies—to provide $5.2 billion to build the Nabucco pipeline. Latest EU planning calls for construction of the 1.1 tcf-capacity Nabucco pipeline to begin in 2012 and for shipments to begin in 2017. In 2011, new higher cost estimates for building the pipeline appeared to place these plans at risk, pushing back shipments until 2019.

As another alternative to natural gas shipments through Turkey, Azerbaijan, Romania, and Georgia signed a memorandum of understanding in April 2010 to transport liquefied natural gas (LNG) from Azerbaijan to the EU through Georgia and Romania. This Azerbaijan-GeorgiaRomania-Interconnection (AGRI) project envisions the construction of a natural gas pipeline from Azerbaijan to the Georgian port of Kalevi, where the natural gas would be liquefied, shipped across the Black Sea, and regasified at the Romanian port of Constanta. This is an unusual proposal to use LNG as the distance across the Black Sea is relatively short—the industry norm for LNG utilization is 1,500 miles. The project output is expected to be 247 bcf per year, with 71 bcf of the natural gas used by Romania and the rest by other EU countries. The presidents of the three countries (and the prime minister of Hungary, which joined the project) met in Baku on September 15, 2010, to sign the Baku Declaration of political support for the project. President Aliyev of Azerbaijan argued that the AGRI project would not make Nabucco less feasible.

Some of the tensions between Turkey and Azerbaijan involving energy issues appeared resolved in June 2010, during President Aliyev's visit to

Turkey, when the two countries signed accords on the sale and transportation of Azerbaijani natural gas to Turkey and to other countries via Turkey. A memorandum of understanding permitting Azerbaijan to conclude direct sales with Greece, Bulgaria, and Syria involving natural gas transiting Turkey was signed. Many observers viewed the MOU as increasing the feasibility of the ITGI and Nabucco pipelines.[26]

In January 2011, President Aliyev and the President of the European Commission, Jose Manuel Barroso, signed a joint declaration committing Azerbaijan to supplying substantial volumes of natural gas over the long term to the European Union. Nonetheless, some analysts raised concerns that there would not be enough Azerbaijani natural gas to fill the ITGI and Nabucco pipelines (deliveries would be 406 bcf per year for ITGI and 158 to 459 bcf per year for Nabucco) and to provide for the proposed AGRI project without a trans-Caspian natural gas pipeline or participation by Iran or Iraq. In 2010, Azerbaijan produced about 530 bcf. Others suggested that Azerbaijan would be able to supply at least most of the needed natural gas for both the ITGI and Nabucco pipelines and the AGRI project, because of recent results from exploratory drilling off the Caspian seacoast.[27]

In September 2011, the Council of the European Union approved opening talks with Azerbaijan and Turkmenistan to facilitate an accord on building a trans-Caspian natural gas pipeline. Such a link would provide added natural gas to ensure adequate supplies for the planned Nabucco and other pipelines. Hailing the decision, EU Energy Commissioner Günther Oettinger stated that "Europe is now speaking with one voice. The trans-Caspian pipeline is a major project in the Southern Corridor to bring new sources of natural gas to Europe. We have the intention of achieving this as soon as possible."[28] The Russian Foreign Ministry denounced the planned talks, and claimed that the Caspian Sea littoral states—Azerbaijan, Iran, Kazakhstan, Russia, and Turkmenistan—had agreed in a declaration issued in October 2007 that decisions regarding the Sea would be adopted by consensus among all the littoral states (Russia itself has violated this provision by agreeing with Kazakhstan and with Azerbaijan on oil and natural gas field development). It also claimed that the proposed pipeline was different from existing sub-sea pipelines in posing an environmental threat.[29]

By the beginning of October 2011, the State Oil Company of Azerbaijan (SOCAR) had received final proposals for pipelines to export natural gas from the second phase development of the Shah Deniz offshore oil and natural gas fields. Proposals were received from consortia backing the ITGI, Nabucco, and Trans Adriatic Pipeline (TAP; from Turkey through Greece, Albania, and the Adriatic Sea to Italy) projects, as well as from BP, which reportedly proposed an 808-mile "South East Europe Pipeline" (SEEP) from western Turkey through Bulgaria, Romania, and Hungary to Austria. A substantial part of the project reportedly would involve building inter-connectors between existing pipelines. A proposal for AGRI was not reported. SOCAR and other members of the Shah Deniz consortium stated that they would decide on a pipeline within several weeks.

On October 25, 2011, Azerbaijan and Turkey announced that they had signed accords on the final terms for the transit of Shah Deniz phase 2 natural gas through the southern corridor. The agreements were signed during President Aliyev's visit to Turkey. They specified that 565-700 bcf of natural gas would transit Turkey, of which 210 bcf would be available for Turkey's domestic use. Another significant accord provided for the possible construction of a new "Trans-Anatolia" natural gas pipeline, so that the natural gas from Shah Deniz Phase 2 would not have to go through the Turkish pipeline system. This pipeline could link to BP's proposed SEEP. In late December 2011, the Azerbaijani and Turkish governments signed a memorandum of understanding on setting up a consortium involving SOCAR, the Turkish state-owned TPAO energy firm, and TPAO's pipeline subsidiary, BOTAS, to construct the Trans-Anatolian Pipeline. SOCAR is designated initially to hold an 80% share in the consortium, although other companies may be invited to join later.

POTENTIAL SOURCES OF ALTERNATIVE SUPPLIES

Global natural gas reserves have increased every year for at least the last three decades, and the advent of shale gas makes the future of natural gas possibly even larger. The US Energy Information Administration (EIA) estimates global natural gas reserves, both conventional and unconventional, at over 6,600 tcf and technically recoverable shale gas resources at about the same, while consumption was about 112 tcf in 2010—or almost 125 years worth of natural gas.[30]

Two regions—Central Asia and North Africa—hold great potential to produce more natural gas than they currently do, and given the proximity of both to Europe (see *Figure 3)* offer possible alternatives to Russian supplies. Central Asia has been a focus of US and European efforts to provide Europe an alternative to Russia for natural gas through the southern corridor. North Africa already has multiple pipelines to Europe and LNG export terminals.

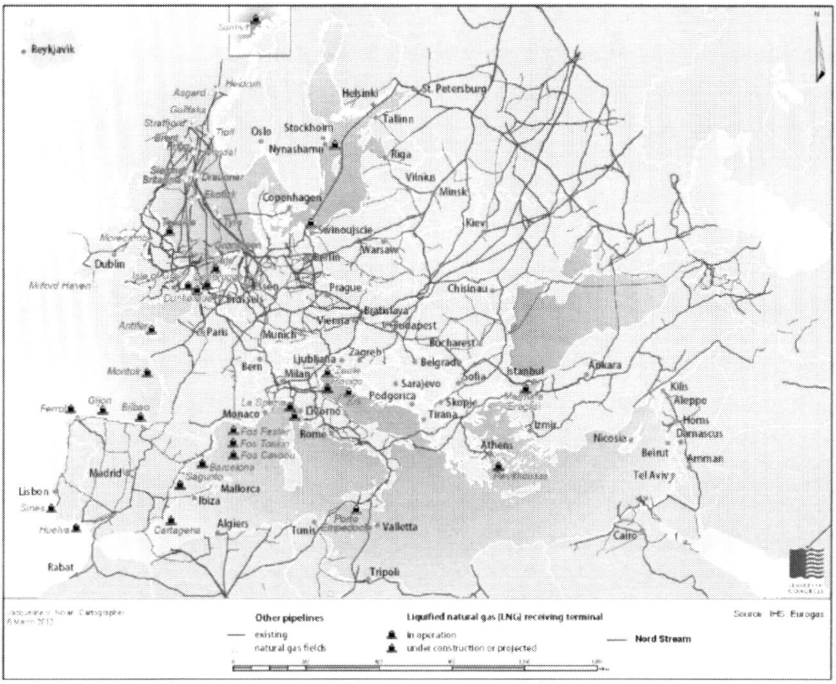

Source: Compiled by the Library of Congress Cartography section.

Figure 3. Select European Natural Gas Infrastructure.

The main issue for this region is whether the MENA nations, with existing reserves and infrastructure, can increase production and delivery of additional supplies to Europe.

There has been tremendous growth in LNG liquefaction over the last few years, mainly in Qatar, and more capacity is projected to be added by industry. Even the United States has multiple proposed LNG liquefaction projects at various stages of regulatory approval. The addition of more liquefaction capacity will provide the EU with other alternative suppliers even though their ability to use LNG is constrained by a lack of infrastructure.

Central Asia and the Caspian Region: The Focus of US Policy[31]

The Caspian region (see *Figure 4*) has emerged as a significant source of natural gas for world markets. The proven natural gas reserves of Azerbaijan, Kazakhstan, Turkmenistan, and Uzbekistan are estimated at over 450 tcf, among the largest in the world and greater than those in Russia (see *Table 3*). The International Energy Agency (IEA) estimates that the Caspian region's proven and recoverable natural gas reserves are about 7% of the world's reserves, but also stresses that further exploration could result in an upward revision of estimated reserves. Nonetheless, the Central Asian states remain geographically isolated from world markets. Natural gas pipelines must be built long distances and must traverse several countries, increasing political and economic risks. Those pipelines which head westward must traverse either the Caspian Sea, where the littoral states continue to argue over its legal status, pass through energy competitors Russia or Iran, or for Azerbaijan, across Turkey.

Source: http://www.david-sadler.org/pages/obx/ukraine.htm.

Figure 4. The Caspian Region.

Asia is a growing prospect for Central Asian natural gas. A natural gas pipeline from Turkmenistan to China exists, but China needs to upgrade its internal supply network to provide natural gas to the coastal industrial areas. Kazakhstan is in discussions with China to export natural gas as well. Turkmen natural gas fields could help meet both Pakistan's and India's growing energy needs and provide significant transit revenues for both Afghanistan and Pakistan."[32] If enough capacity is constructed to China and other parts of Asia, future supplies to Europe may be moot, which would benefit Russia.

Table 3. Key Central Asian Natural Gas Data, 2010
Units = trillion cubic feet (tcf)

	Reserves	Production	Exports to EU
Azerbaijan	44.9	0.5	0.0[a]
Kazakhstan	65.2	1.2	0.0
Turkmenistan	283.6	1.5	0.0
Uzbekistan	55.1	2.1	0.0
TOTAL	448.8	5.3	0.0

Source: BP Statistical Review of World Energy 2011, pp. 20, 22, and 28.

[a]. Azerbaijan does export natural gas to Turkey, which then sends some of it to Greece.

Azerbaijan: The EU's Best Hope For New Natural Gas Supplies?[33]

US administrations have contested that exports from Azerbaijan could boost energy security for European customers currently relying more on Russia. According to Ambassador Morningstar, Azerbaijani natural gas "is absolutely essential to the development of the Southern Corridor," and will be able to supply at least some if not most of the needed natural gas for both the proposed Interconnector-Turkey-Greece-Italy (ITGI) natural gas pipeline and the first phase of the Nabucco pipeline, if built.[34] In March 2007, Azerbaijan and the United States signed a memorandum of understanding on energy cooperation that called for discussions on the proposed ITGI and Nabucco natural gas pipelines. In August 2007, the US Trade & Development Agency granted Azerbaijan $1.7 million to fund feasibility studies on building both a natural gas and an oil pipeline across the Caspian Sea to link Central Asia to the SCP pipeline and the Baku Tbilisi Ceyhan pipeline.

Of importance to US foreign policy is Azerbaijan's relationship with Iran. At the end of 2005, Azerbaijan began sending about 7 billion cubic feet of natural gas per year through a section of Soviet-era pipeline to the Iranian border at Astara, partly in exchange for Iranian natural gas shipments to Azerbaijan's Nakhichevan exclave.[35] On November 11, 2009, Azerbaijan signed an accord with Iran to supply 17.7 bcf of natural gas annually through the pipeline. These natural gas supplies could increase in coming years.

Kazakhstan: Natural Gas Is Second to Oil[36]

Most natural gas production in Kazakhstan has been associated with the development of oil fields, and most of the natural gas has been re-injected into the fields. Natural gas is mostly produced in the northwestern part of the country, while population centers in the eastern and southern parts are dependent on natural gas imported from Uzbekistan. In 2009, Kazakhstan became a net natural gas exporter. According to the BP Statistical Review, Kazakhstan exported about 424 billion cubic feet (bcf) of natural gas from its western fields to Russia in 2010. In December 2007, Kazakhstan, Turkmenistan, and Russia signed an agreement to renovate a branch of the Central Asia-Center Pipeline supplying natural gas to Russia and to build a new Caspian Coastal Pipeline, but these plans have been delayed by Turkmenistan's intentions to diversify its export routes away from Russia and by reduced natural gas demand by Russia. Kazakhstan nonetheless plans to boost its natural gas exports in coming years to Russia and China.

Until recently, US foreign direct investment (FDI) played a dominant role in the development of Kazakhstani oil and natural gas resources, amounting to about $29 billion in Kazakhstan (over one-third of all FDI in the country) from 1993-2009.[37] According to some reports, China provided about $13 billion in investments and loans to Kazakhstan's energy sector in 2009, highlighting its rising energy influence. Some US energy firms and other private foreign investors have become discouraged in recent months by harsher Kazakh government terms, taxes, and fines that some allege reflect corruption within the ruling elite.

At the end of October 2008, China and Kazakhstan signed a framework agreement on constructing a natural gas pipeline from Beyneu, north of the Aral Sea, southeastward to Shymkent, where it will connect with the Central Asia-China Gas Pipeline. The 932-mile Beyneu-Shymkent Pipeline link is planned initially to supply 176.6 bcf to southeastern Kazakhstan and 176.6 bcf to China. Pipeline construction began in September 2011 and is expected to be completed by 2015.

Kazakh officials have appeared to make contradictory statements about providing natural gas for the prospective Nabucco pipeline. Kazakhstan's Deputy Energy and Mineral Resources Minister Aset Magaulov stated at a Euro-Atlantic Partnership Council Security Forum in June 2009 that Kazakhstan would not have a surplus of natural gas that it could send through the Nabucco pipeline.[38] Later that year, President Nazarbayev appeared to support the possible transit of Kazakh natural gas through Turkey. In late October 2009, however, the Kazakh Ministry of Energy reiterated that "the main problem for our country [regarding the supply of natural gas to Nabucco] is the limited availability of gas" because of existing contracts for projected natural gas production. It suggested that Kazakhstan might be a potential supplier for Nabucco if natural gas production exceeds expectations, but that Kazakhstan could not transport any natural gas via Nabucco until the legal status of the Caspian Sea was resolved, which would permit building a connection to Nabucco.[39] In early October 2011 Minister of Oil and Gas Sauat Mynabyev stated that "we do not have available resources for the gas pipeline yet."[40]

Turkmenistan: European Orientation?[41]

As shown in *Table 3*, Turkmenistan holds the largest natural gas reserves in Central Asia. A significant quantity of Turkmen natural gas production already flows to Europe via Russia. However, Turkmenistan's drive for alternative export routes for its natural gas has pitted it against some of the other Caspian countries. In September 2011, the Council of the EU approved opening talks with Azerbaijan and Turkmenistan to facilitate an accord on building a trans-Caspian natural gas pipeline. Russia and Iran oppose the building of trans-Caspian pipelines, claiming that the delineation of Caspian Sea borders and the use and protection of maritime resources must first be worked out by the littoral states. Many observers view such objections as partly driven by the status of Russia and Iran as natural gas producers in competition with Turkmenistan. Russia, in particular, appears to want to maintain its role as a major importer of Turkmen natural gas and to prevent it from competing directly with Russian natural gas exports to the EU. Turkmenistan's claims against Azerbaijan regarding some offshore oil and natural gas fields also have stymied a formal agreement on a trans-Caspian pipeline between the two countries. In mid-October 2011, Russian President Medvedev warned again that all the littoral states would need to agree to a trans-Caspian pipeline. The Turkmen Foreign Ministry retorted by terming this stance "counterproductive" to Turkmen-Russian relations. The Foreign Ministry pointed out that several bilateral agreements on sea use had been concluded by Russia and others, and repeated Turkmenistan's argument that it similarly could reach an agreement with Azerbaijan on a pipeline.

Despite Turkmenistan's desire to export more of its gas, thus far, its orientation seems to be toward the east and not yet toward Europe. Turkmenistan has been seeking alternatives to pipeline routes through Russia for some time. Since December 1997 Turkmenistan has opened two pipelines to Iran doubling Turkmenistan's export capacity to Iran to about 700 bcf per year.[42]

In April 2006, Turkmenistan and China signed a framework agreement calling for Chinese investment in developing natural gas fields in

Turkmenistan and in building a natural gas pipeline through Uzbekistan and Kazakhstan to China, which is in operation. Finally, Turkmen President Berdimuhamedow also has revived his predecessor's proposal to build a natural gas pipeline through Afghanistan to Pakistan and India (TAPI).

Uzbekistan: A Sleeping Natural Gas Giant?[43]

Uzbekistan mostly uses its natural gas production domestically and is self-sufficient. It has, however, used its network of Soviet-era natural gas pipelines to export some natural gas to Russia and to other Central Asian states (Kazakhstan, Kyrgyzstan, and Tajikistan). Uzbekistan appears to have sufficient gas reserves to become a potential supplier of some gas to Europe if its infrastructure development begins to look westward.

However, Uzbekistan has been largely closed to Western energy investment, although efforts to attract international energy firms appeared to increase in 2010-2011. Russian firms Gazprom and Lukoil are the largest investors in Uzbek natural gas development and production and seem through their policies to want to keep Uzbek natural gas from competing with other Russian natural gas being supplied to Europe. In 2005, the Central Asia-China Pipeline (CNPC) and Uzbekistan's state-owned Uzbekneftegaz announced that they would form a joint venture to develop oil and natural gas resources. In 2007, Uzbekistan and China signed an agreement on building a 326-mile section of the CNPC pipeline, and a construction and operation joint venture between Uzbekneftegaz and CNPC, Asia Trans Gas, began construction in 2008. Uzbekistan also has signed a framework agreement to eventually supply 353 bcf of natural gas per year through the pipeline. A production sharing consortium composed of Uzbekneftegaz, Lukoil, the Korea National Oil Corporation, and CNPC is exploring for natural gas in the Aral Sea region.

North Africa: Transitions May Bring Opportunities[44]

To date, US energy strategy towards Europe has not focused on North Africa as a counter balance to Russian natural gas supplies. Nevertheless, the Arab Spring may have created an opportunity, albeit with potential problems, to increase exports from the region. Taken as a whole, the three main existing suppliers to Europe in the region—Algeria, Egypt, and Libya—already supply natural gas to Europe by both pipeline and LNG (see *Table 4)* and hold tremendous natural gas resources that could be further developed. Collectively, the three countries supply about 60% of what Russia supplies, of which Algeria is the source for almost 80%. Difficult business environments and domestic demand, prompted by subsidies for natural gas consumption, have limited development of each country's natural gas resources. Regime changes in Egypt and Libya pose an opportunity for each to change its policies to promote expanded development of natural gas resources. At the same time, political and economic uncertainty could continue to characterize the situation in both countries in the short- to medium-term.

Table 4. Key North African Natural Gas Data, 2010
Units = trillion cubic feet (tcf)

	Reserves	Production	Exports to EU
Algeria	159.1	2.8	1.8
Egypt	78.0	2.2	0.2
Libya	54.7	0.6	0.3
TOTAL	291.8	5.6	2.3

Source: BP Statistical Review of World Energy 2011, pp. 20, 22, and 28.

Algeria: Unconventional Resources May Be Its Future[45]

According to a study by the US Energy Information Administration (EIA), Algeria may hold shale gas resources much greater than its conventional reserves, which are substantial. Depending upon the development of its unconventional natural gas resources and its conventional resources, Algeria could become a more significant natural gas producer and exporter. However, a difficult business environment may continue to limit its potential.

A 2005 hydrocarbon law diminished the monopoly of the state energy company, Sonatrach, opening the sector for private and foreign investment. A 2006 law, however, required international companies to give Sonatrach a 51% stake in new oil, natural gas, and related transport projects. Additional foreign investment rules were enacted in the Complementary Finance Law (CFL) of 2009, which restricted imports and foreign investment. These measures require 51% Algerian ownership of new foreign investment. Further, the 2010 CFL, effective as of September 2010, requires foreign bidders who win construction contracts to invest in a joint venture with a local partner.[46] Such changes have prompted foreign investors, including US and European businesses and governments, to appeal for greater stability of laws in Algeria, and may have contributed to a reported slowing of foreign investment in exploration and production.[47]

Algerian natural gas production and exports have declined since 2005 when it produced over 3.1 tcf and exported more than 2.2 tcf. In 2010, Algeria produced 2.8 tcf and exported 2.0 tcf, with 1.8 tcf going to the EU. In 2005, Algeria's energy minister announced ambitious plans to increase production and export, with a goal of reaching 4.0 tcf of production and 3.5 tcf of exports by 2015. These targets are not on track to be achieved, and the country has changed its focus to preserving its resource base and not expanding production as quickly.

Nevertheless, Algeria continues to expand its connections to Europe. In 2011, a consortium led by Sonatrach opened the Medgaz natural gas pipeline. The new pipeline runs directly from Algeria's Beni Saf port to Spain's Perdigal Beach. The initial capacity of the line is approximately 280 bcf per year. Despite this new addition, Algerian exports to Spain do not have much impact on the rest of Europe, as the interconnection between Spain and France is limited. In addition to Medgaz, Algeria exports natural gas to Europe via the 425 bcf Maghreb-Europe pipeline to Spain and the 230 bcf Trans-Mediterranean pipeline to Italy. Algeria has also announced plans to expand its LNG export capacity.

Egypt: In Need of a Reorganization of Its Natural Gas Sector[48]

Since 2005, demand for natural gas in Egypt has been on the rise, increasing almost 43% over the time period. Although production has grown as well, the subsidy-driven demand has hindered the government in offering attractive terms for international companies to continue developing Egypt's resources. Additionally, much of Egypt's remaining natural gas is in difficult-to-access, high-cost areas, which contributes to the lack of interest by many international natural gas companies. That said, BP signed a deal in 2010 that was substantially higher than previous contract terms.

Since the resignation of former Egyptian President Hosni Mubarak in February 2011, Egypt's natural gas infrastructure in the Sinai Peninsula has been attacked at least ten times by either disaffected Bedouin Arabs living in the Sinai or terrorist groups with camps in the peninsula. These attacks have disrupted gas shipments via two separate pipelines converging at El Arish to both Israel and Jordan. No group has claimed responsibility for the attacks, and the Egyptian authorities have struggled to protect infrastructure in the demilitarized Sinai Peninsula. Israel generates 40% of its electricity using natural gas, and Egypt provides 43% of its supplies. Egyptian natural gas supplies 80% of Jordan's power generation needs.

Egyptian exports to the EU, which are solely in the form of LNG, dropped by almost 35% in 2010. The Arab Gas Pipeline from Egypt to Jordan, Lebanon, and Syria has been planned to extend to Turkey in order to move Egyptian natural gas to Europe, but given the issues surrounding Egypt's natural gas sector this is doubtful. Production last year also fell for the first time in over a decade. With domestic consumption likely to continue increasing and production probably continuing to decline, exports are not likely to increase for some time. Depending upon the orientation of a new government, if it promotes western investment in Egypt's energy sector, and the government addresses its natural gas subsidies, this deterioration of Egypt's natural gas sector could be reversed.

Libya: Tremendous Potential[49]

Libya may have the greatest potential to increase natural gas exports to Europe once a new regime is established and possibly a new state oil and natural gas company in a post-Qadhafi Libya. The civil war halted natural gas production, but production has since resumed and appears to be recovering quicker than most analysts had forecast.

Libya has one natural gas pipeline to Europe, Greenstream, which was closed during the recent unrest, as well as an LNG export terminal. Italy received almost 97% of Libya's natural gas exports in 2010, while Libya provided approximately 10% of Italy's natural gas imports. The pipeline was operating close to its capacity, so to significantly increase pipeline exports would require an additional pipeline to be constructed or the existing pipeline to be expanded. Libya's LNG exports were mostly to Spain and minimal in 2010. LNG exports were approximately 10% of the capacity of Libya's LNG facility.

Libya's natural gas production has almost tripled to about 560 bcf since 2003, in part to meet pipeline exports which started in 2004. However, domestic consumption, particularly for electric power generation, could increase Libya's consumption of natural gas, which has been stable over the past decade according to EIA.[50]

Liquefied Natural Gas Imports

One of the most important developments for Europe has been the growing availability of natural gas in liquefied form (LNG). LNG represents about 15% of European natural gas imports. Spain, where 65% of natural gas imports are

LNG, leads Europe in LNG imports, followed by Portugal and France. However, as noted earlier, the interconnection between Spain and France is small and does not allow Europe to take advantage of Spain's import capacity for LNG or pipeline natural gas.

The principal suppliers of LNG to Europe include Algeria, Egypt, Oman, and Qatar. Algeria is the world's third largest exporter of LNG, with almost all of its natural gas going to Europe.

Countries such as Poland and Estonia have also begun the process of building large LNG import terminals at their Baltic Sea ports that will enable LNG to be distributed throughout northern and eastern Europe. Qatar, which supplied about 10% of the EU's imports, also owns multiple LNG import terminals in Europe.

Table 5. EU LNG Import Capacity

	Number of Facilities	Capacity (bcf)
Belgium	1	9.0
France	3	23.8
Greece	1	5.3
Italy	2	11.0
Netherlands	1	12.0
Portugal	1	6.5
Spain	6	60.1
United Kingdom	4	51.1
	19	178.7

Source: Gas Infrastructure Europe, http://www.gie.eu.com/index.php/maps-data/lng-map.

Possible US LNG Exports: Pricing Not Volumes May Be Key

Proposed US LNG export projects, if all were constructed today, would make the United States the second largest LNG exporter behind Qatar. The proposed projects are at various stages in the regulatory approval process. Nevertheless, analysts have already begun speculating on what a significant increase in US LNG exports would mean to natural gas markets, especially to European markets. Any volumes of LNG from the United States would benefit the market, including Europe, by offering a new supplier to consumers. For parts of Europe, especially the Baltic region and Central Europe, where the United States enjoys strong and friendly relations, any decision to export US

LNG to that region would be welcomed as a potential offset to their dependence on Russian gas.

However, the bigger effect of US entry into global LNG sales may be on pricing rather than supplies. The United States is one of the few countries that does not link its natural gas price to the price of oil and therefore may add to the pressure to delink the two commodities. Most natural gas sold in the world, by pipeline or as LNG, is sold under long-term contracts and indexed to the price of oil. Historically, the two commodities competed more directly in markets than they do today.

More Distant Alternatives

Eastern Mediterranean: A Recent Development

Although too early to tell and years from production for export, recent announcements of natural gas discoveries in the eastern Mediterranean by Israel and Cyprus may open a new source of European natural gas. Initial estimates pose a scenario in which Israel and Cyprus could become natural gas exporters, with Europe as the largest nearby market a likely recipient. Cyprus, which is an EU member, currently does not consume any natural gas in its economy and would require much infrastructure to do so. However, both Israel and the US energy company Noble Energy, which is conducting the drilling, have raised the potential to help Cyprus build natural gas facilities for both domestic consumption and export. Additionally, other countries in the region, including Turkey, may begin exploration efforts that could increase the amount of natural gas produced in the region.

The Arctic Region and Players

Norway is not a member of the EU, but is the eighth-largest natural gas producer in the world and second-largest exporter of natural gas to the EU, behind Russia.

The North Sea holds the majority of Norway's natural gas reserves, but there are also significant quantities in the Norwegian and Barents Seas. The United States Geological Survey has estimated that almost 25% of the globe's yet-to-be-discovered natural gas resources are located in the Arctic region and last year Norway and Russia reached agreement on Arctic energy exploration issues. Norway's Snohvit natural gas field along with Russia's field at Shtockman, in which Norway is an investor and development partner, promises to make the Barents Sea a new European energy region.

Potential Development of Alternative Sources in Europe

In addition to solidifying other sources of energy supply from other regions, experts point to several additional factors that could decrease European dependence on Russian resources. The development of previously difficult-to-develop "unconventional" natural gas deposits, including shale gas, in Europe and elsewhere could diversify supplies and keep prices down. EIA assessed the EU's technically recoverable shale gas resources at almost 500 tcf, more than 25 years of supply at current consumption levels.[51] The growth of the spot market for natural gas and the development of liquefied natural gas infrastructure in Europe could also help diversify supplies as well as reduce dependence on Russian-controlled pipelines. Finally, developing alternative energy sources within Europe, in particular, hydropower, energy from the seas, biomass, wind power, solar energy, and geothermal energy could all contribute to further diversification of Europe's energy supply, reducing overall natural gas demand.

PROSPECTS FOR DIVERSIFICATION

There are many alternatives to Russian natural gas for Europe to chose from, but it would be difficult, if not impractical, for Europe to consider replacing all Russian natural gas imports. There is also inertia on the part of some EU countries and companies regarding the status quo. Some of Europe's larger natural gas companies have huge financial interests in maintaining Russian supplies and do not see a problem in depending so much on one country. It is important to keep in mind that not only does Russia hold the largest supplies of natural gas globally, but already has significant infrastructure connecting its resources to Europe, while some of the alternatives remain constrained. A major test for the EU in developing a more coherent energy policy for Europe will be how to balance these views with those of other member states that are more dependent on Russian energy and are concerned by the political leverage Russia could exert on parts of Europe if no alternatives are found to alleviate at least some of that dependence.

Although supplying natural gas to Europe from Central Asia, particularly through the proposed Nabucco pipeline, has been a goal of multiple US administrations and the EU, it is far from being achieved in volumes significant to counter Russian exports. Recent statements by Ambassador Morningstar suggesting that the US administration would accept any economically viable pipeline could indicate waning US interest in the Nabucco project. In addition, given the interest in combating climate change both in Europe and in some quarters of the United States, many analysts believe that having Caspian natural gas go to China instead, where a pipeline already exists, could help that country decrease its carbon dioxide and other greenhouse gas emissions by, for instance, limiting its use of coal in China's electric power sector.

In North Africa, ongoing governmental transitions in Libya and Egypt are a key factor for natural gas development. In January 2012, Egypt held its first parliamentary elections since the ouster of President Hosni Mubarak. Libya is preparing for elections in June 2012 to replace the interim government in place since Muammar al Qadhafi's government was toppled in 2011. The type and character of the new governments will have an impact on natural gas development in each country as the energy sectors in each country appear to offer the fastest potential source of economic growth and income for the new governments. Both countries have large natural gas resources, but historical political constraints have limited the development of these resources.

The United States and Europe are in a position to aid both countries in reforming their regulatory regimes governing natural gas development as well as establishing oversight by nongovernmental organizations and their respective parliaments. And both the US and European energy companies seem eager to help further develop energy infrastructure and production in both countries. Redirecting US and European efforts from Central Asia to MENA—especially Libya and Egypt—as an alternative to Russian natural gas supplies could improve the chances of more natural gas reaching Europe in the short run. Meanwhile, new discoveries in the eastern Mediterranean pose a potential new source of European natural gas. However, neither Israel nor Cyprus have any experience in developing large scale natural gas projects. Both countries could benefit from the US and European experience in developing their resources, both on a federal and state level.

End Notes

[1] US Energy Information Administration, World Shale Gas Resources: An Initial Assessment of 14 Regions Outside the United States, Washington, DC, April 5, 2011, p. 4, http://www.eia.gov/analysis/studies/worldshalegas/.

[2] The southern corridor refers to the area south of the Black Sea and into southern Europe.

[3] For additional information on US natural gas exports see CRS Report R42074, US Natural Gas Exports: New Opportunities, Uncertain Outcomes, by Michael Ratner, Paul W. Parfomak, and Linda Luther.

[4] The White House, Vice President's Remarks at the Vilnius Conference, May 4, 2006, http://www.whitehouse.gov.

[5] US House of Representatives, Committee on Foreign Affairs, Subcommittee on Europe and Eurasia, Hearing on European and Eurasian Energy: Developing Capabilities for Security and Prosperity, Testimony of Ambassador Richard L. Morningstar, Special Envoy for Eurasian Energy, June 2, 2011. For additional information on Turkey see CRS Report R41368, Turkey: Background and US Relations, by Jim Zanotti.

[6] At the time of the testimony, the South East Europe Pipeline (SEEP) and Trans-Anatolia projects did not exist.

[7] US House of Representatives, Committee on Foreign Affairs, Subcommittee on Europe and Eurasia, Hearing on European and Eurasian Energy: Developing Capabilities for Security and Prosperity, Testimony of Ambassador Richard L. Morningstar, Special Envoy for Eurasian Energy, June 2, 2011.

[8] US Embassy, Baku, Azerbaijan, Office of Public Affairs, Media Advisory: Special Envoy for Eurasian Energy Supports Southern Corridor, November 17, 2011.

[9] Office of the Press Secretary, "Remarks by the President on the Middle East and North Africa," The White House, State Department, Washington, DC, May 19, 2011.

[10] E.g., see Carnegie Endowment for International Peace, "Lieberman Delivers Remarks on Democratic Transition in Egypt," July 22, 2011, http://carnegieendowment. org/files/ Lieberman_Prepared_Remarks.pdf; John McCain, Lindsey Graham, Mark Kirk, and Marco Rubio, "The Promise of a Pro-American Libya," Wall Street Journal, October 7, 2011. On November 18, 2011, Representative Dreier introduced a resolution, co-sponsored by Representative Meeks, that calls for the United States to initiate FTA negotiations with Egypt (H.Res. 472).

[11] For more information, see (forthcoming) CRS Report, US Trade and Investment in the Middle East and North Africa: Overview and Issues for Congress, coordinated by Rebecca M. Nelson.

[12] European Commission, Market Observatory for Energy, Key Figures, June 2011. http://ec. europa.eu/energy/ observatory/eu_27_info/doc/key_figures.pdf.

[13] Eurogas, Long Term Outlook for Gas Demand and Supply 2007-2030, June 5, 2010, p. 5, http://www.eurogas.org/ uploaded/Eurogas%20LT%20Outlook%202007-2030_Final_ 251 110.pdf.

[14] Comments provided through discussions with representatives of several European member states.

[15] Denis Pinchuk, "Gazprom Drops Austria from S. Stream Gas Route—Source," Reuters, December 14, 2011.

[16] "Internal energy market in doubt as 18 states face court," Euractiv, October 3, 2011.

[17] Tim Webb, "European gas pipeline costs double," The Guardian, February 20, 2011; the Nabucco consortium is made up of six European energy companies: OMV of Austria; MOL of Hungary; Transgaz of Romania; Bulgarian Energy Holding of Bulgaria; BOTAS of Turkey; and RWE of Germany.

[18] For additional information on Russia see CRS Report RL33407, Russian Political, Economic, and Security Issues and US Interests, coordinated by Jim Nichol.

[19] Russia also supplies the EU with about 27% of its oil imports, 24% of its coal imports, 30% of it uranium imports, and is the third largest supplier of electricity imports, but these fuel sources are beyond the scope of this report.

[20] The Commonwealth of Independent States (CIS) includes Armenia, Azerbaijan, Belarus, Kazakhstan, Kyrgystan, Moldova, Russia, Tajikistan, and Uzbekistan with Turkmenistan and Ukraine having unofficial status. Georgia withdrew from the CIS in 2009.

[21] Sergey Paltsev, "Russian Natural Gas Export Potential Up to 2050," MIT Center for Energy and Environmental Policy Research, July 2011.

[22] The text of the National Security Strategy can be found at the website of the Russian National Security Council at http://www.scrf.gov.ru/documents/99.html.

[23] For more information on Russia's official energy strategy, see Energy Strategy of Russia for the Period up to 2030, at http://www.energystrategy.ru/projects/docs/ES-2030_(Eng).pdf.

[24] Luke Harding, "Ukraine extends lease for Russia's Black Sea Fleet," The Guardian, April 21, 2010, http://www.guardian.co.uk/world/2010/apr/21/ukraine-black-sea-fleet-russia.

[25] "Value of Ukraine Gas Transportation System Could Fall Seriously - Miller," Interfax, December 31, 2011.

[26] "Azerbaijan to Move Quickly to Negotiate Sales of Gas to Europe," Oil Daily, June 14, 2010.

[27] Eric Watkins, "New USGS Report Confirms Big Caspian Stakes," Oil & Gas Journal, January 3, 2011.

[28] European Commission, Press Release: EU Starts Negotiations on Caspian Pipeline to Bring Gas to Europe, September 12, 2011.

[29] "Moscow 'regrets' EU decision on trans-Caspian gas pipeline," RIA Novosti, September 13, 2011, Web Edition.

[30] US Energy Information Administration, World Shale Gas Resources: An Initial Assessment of 14 Regions Outside the United States, Washington, DC, April 5, 2011, p. 4, http://www.eia.gov/analysis/studies/worldshalegas/. Reserves and resources are not the same in the energy industry. Reserves are considered a subset of resources as they indicate that a resource is producible using today's technology at today's prices.

[31] For additional information on Central Asia see CRS Report RL33458, Central Asia: Regional Developments and Implications for US Interests, by Jim Nichol.

[32] US Department of State, Secretary Clinton Co-Chairs the New Silk Road Ministerial Meeting, DipNote, September 23, 2011; Fact Sheet on New Silk Road Ministerial, September 22, 2011. See also US Department of State, Remarks, Robert D. Hormats, Under Secretary for Economic, Energy and Agricultural Affairs, Address to the SAIS Central Asia-Caucasus Institute and CSIS Forum, September 29, 2011; William J. Burns, Deputy Secretary of State, Remarks at Istanbul Conference for Afghanistan, November 2, 2011.

[33] For additional information on Azerbaijan see CRS Report 97-522, Azerbaijan: Recent Developments and US Interests, by Jim Nichol.

[34] US Embassy, Baku, Media Advisory, Ambassador Richard Morningstar, Special Envoy for Eurasian Energy: Speech to Plenary Session of Caspian Oil and Gas Conference, June 8, 2011.

[35] The Nakhichevan exclave is Azerbaijani territory that is situated between the Armenian controlled area of NagornoKarabagh and Armenia proper.

[36] For additional information on Kazakhstan see CRS Report 97-1058, Kazakhstan: Recent Developments and US Interests, by Jim Nichol.

[37] US Department of State. Robert O. Blake, Jr., Assistant Secretary, Bureau of South and Central Asian Affairs: Remarks Before the Washington International Business Council, February 24, 2010.

[38] ITAR-TASS, June 25, 2009.

[39] ITAR-TASS, October 31, 2009.

[40] Interfax, October 6, 2011.

[41] For additional information on Turkmenistan see CRS Report 97-1055, Turkmenistan: Recent Developments and US Interests, by Jim Nichol.

[42] According to the BP Statistical Review, actual Turkmen natural gas exports were about 230 bcf to Iran in 2010.

[43] For additional information on Uzbekistan see CRS Report RS21238, Uzbekistan: Recent Developments and US Interests, by Jim Nichol.

[44] For additional information on North African energy see CRS Report R41683, Middle East and North Africa Unrest: Implications for Oil and Natural Gas Markets, by Michael Ratner and Neelesh Nerurkar.

[45] For additional information on current events in Algeria, see CRS Report RS21532, Algeria: Current Issues, by Alexis Arieff.

[46] State Department, Bureau of Economic, Energy and Business Affairs, "2011 Investment Climate Statement— Algeria," March 2011.

[47] Samuel Ciszuk, "Abysmal Licensing Round Result in Algeria Confirms Energy Sector Malaise," IHS Global Insight, March 18, 2011; Africa Energy Intelligence, "Oil Companies Steer Clear of Algeria Due to Windfall Tax, Few Bid for Licenses," March 23, 2011.

[48] For additional information on Egypt's energy sector see CRS Report R41632, Implications of Egypt's Turmoil on Global Oil and Natural Gas Supply, by Michael Ratner, and for additional information on current events in Egypt see CRS Report RL33003, Egypt in Transition, by Jeremy M. Sharp.

[49] For additional information on current events in Libya see CRS Report RL33142, Libya: Transition and US Policy, by Christopher M. Blanchard.

[50] US Energy Information Administration, Country Analysis Briefs - Libya, Washington, DC, February 2011, http://www.eia.gov/EMEU/cabs/Libya/pdf.pdf.

[51] US Energy Information Administration, World Shale Gas Resources: An Initial Assessment of 14 Regions Outside the United States, Washington, DC, April 2011, p. 4, http://www.eia.gov/analysis/studies/worldshalegas/.

In: European Energy Security Options ... ISBN: 978-1-62081-814-5
Editors: R. E. Phillips and J. B. Cook

Chapter 2

STATEMENT OF THE HONORABLE GREGORY W. MEEKS. HEARING ON "EUROPEAN AND EURASIAN ENERGY: DEVELOPING CAPABILITIES FOR SECURITY AND PROSPERITY"*

Thank you Chairman Burton for calling this hearing on European and Eurasian energy issues.

Energy security issues in Europe and Eurasia are directly connected to prosperity, economic growth, regional integration, stability and peace, and therefore of importance to the entire region as well as to the United States.

Europe and Eurasia's energy situation presents a complex picture of unevenly distributed energy scenarios. Collectively, EU member states import over half their energy supply, including approximately 84% of their oil and 62% of their natural gas. Some EU countries are almost entirely reliant on outside providers of energy, while others are largely self-sufficient or even net energy exporters.

The EU's neighbor, Russia, possesses the largest natural gas reserves in the world and also large oil reserves. The Central Asia, Caspian and Black Sea regions offer opportunities for much needed diversification strategies, if successfully connected to European markets, but uncertainties surround the various pipeline projects currently under discussion. The Middle East and North Africa also provide substantial oil and gas resources to Europe, and has

* This is an edited, reformatted and augmented version of testimony presented June 2, 2011 before the House Foreign Affairs Committee.

the potential to increase these amounts if a more stable and open investment climate can be achieved. Energy exploration is increasing in the Arctic and could potentially provide vast amounts of untapped resources for both European and global markets.

Overall, Europeans are pursuing a number of strategies to address their future energy needs, including development of alternative energy sources, market liberalization, research and investment in renewable energy, increased efficiency, and ambitious policy goals.

Generally, Europeans are ahead of the curve in two ways: utilizing renewable energy and reducing energy consumption. The United States can learn from both strategies. Consider, for example, that according to the World Bank, Americans use almost twice as much energy per capita as many of the European countries that we would normally compare ourselves with. Simultaneously, the European Union has a target of achieving 2 0% of their energy supply from renewables by 2 02 0. Several countries are already far ahead of this goal, utilizing hydro, solar, and wind power aggressively to diversify their national energy markets. The percentage of renewables in the US energy mix is around 11%, for comparison.

The EU has also led cooperation to create interconnectedness across the European continent, in order to ensure energy security, prevent dependency on single suppliers and prevent supply disruptions. Despite all these efforts, geography cannot easily be changed, and in many ways, Europe's destiny is to have Russia as an "indispensable partner for European energy security", in the words of EU Commissioner for Energy, Gunter Oettinger. I look forward to carefully examining this relationship today.

Russia's relationship with Europe — and with the United States — is truly multi-faceted. On one hand, Russia has used its dominant role as energy supplier towards political leverage in Central and Eastern Europe, the Baltics and the Balkans. On the other, Russia has proven a valuable partner in bringing peace and stability to Afghanistan by facilitating the Northern Distribution Network supply route, engaged with the United States to reduce our nuclear arsenals and joined the US and EU in approving UN sanctions towards Iran's nuclear weapons program.

This leads me to wonder whether the European dependence on Russian oil and gas is sometimes exaggerated. The increased availability of shale gas and liquefied natural gas resources combined with increased use of renewable and efficiency gains, tougher implementation of EU competition and anti=monopoly laws and the prospects of Caspian, North African and Arctic energy resources suggest that Europe may be in better shape over the long term than we give them credit for. I would be interested in our witnesses' assessments of whether this is the case, and whether the recent announcement by Germany's Chancellor Angela Merkel to phase out nuclear power in Germany by 2 022 will change the overall scenario.

These developments are taking place as the Middle East, North Africa and potentially parts of Central Asia are changing before our eyes. I know that Ambassador Morningstar has worked tirelessly to pursue US diplomatic goals with the European Union, Turkey, Iraq, Azerbaijan and a host of Central Asian countries to fulfill American goals of developing new resources, assisting European energy security and helping Caspian and Central Asian countries connect their resources to global markets. This engagement is particularly important, as Central Asia sometimes finds itself beyond the reach of the traditional transatlantic and European institutions, although they certainly play a key role in any discussion relating to European and Eurasian energy security issues.

Again, I want to thank Chairman Burton for holding this hearing, and I look forward to hearing the testimony from our witnesses.

In: European Energy Security Options ... ISBN: 978-1-62081-814-5
Editors: R. E. Phillips and J. B. Cook

Chapter 3

TESTIMONY OF AMBASSADOR RICHARD L. MORNINGSTAR, SPECIAL ENVOY FOR EURASIAN ENERGY. HEARING ON "EUROPEAN AND EURASIAN ENERGY: DEVELOPING CAPABILITIES FOR SECURITY AND PROSPERITY"*

Chairman Burton, Ranking Member Meeks, Members of the Committee:

It is a great pleasure to be here today to talk about one of the important pillars of US diplomacy, energy security. I'd like to thank the Committee for inviting me to talk to you about what the US is doing with regard to energy security. We'll examine the overall US Eurasian energy strategy and then discuss the specific areas where we are implementing it.

WHAT IS THE US STRATEGY?

There are three main components of our Eurasian energy strategy. *First*, we want to encourage the development of new oil and gas resources and also promote efficiency and conservation in the use of all energy resources. Because there is a world market for oil, new production contributes to meeting growing demand anywhere in the world, including in the United States. When

* This is an edited, reformatted and augmented version of testimony presented June 2, 2011 before the House Foreign Affairs Committee.

we are talking about new natural gas production in the Caspian region, it is unlikely that any of that gas will reach the US, but it is still important because it will add to international gas supply. Additional supply in one place naturally frees up supply in another. As the market for liquefied natural gas continues to grow, we can start to think about gas moving around markets in much the same way oil does.

Second, we want to assist Europe in its quest for energy security. With the combination of goods and services, the EU27 and the US account for the largest bilateral trade and investment relationship in the world. Europe is our partner on any number of global issues from Afghanistan to Libya to the Middle East, from human rights to free trade. We have an interest in an economically strong Europe. Of course, Europe is composed of many different states and energy security is a more pressing issue to some than to others. Some countries in Europe do not have a diverse energy mix and depend largely on a single supplier and transport route. When that route is disrupted, as we witnessed in January 2009, the consequences can be severe.

The populations of Bulgaria and Serbia and others who suffered in the cold can attest to that. So our aim is to encourage the development of a balanced and diverse energy strategy with multiple energy sources with multiple routes to market. This approach furthers competitive, efficient markets and the best prices for consumers.

Third, we want to help Caspian and Central Asian countries find new routes to market. We want to help foster economic growth and prosperity in these countries. By expanding export routes, they can increase competition for their resources, demand a fair price, and create strong links to the global economy. These countries should also be able to make their own independent choices regarding how they deal with energy resources.

How Will We Achieve Our Energy Security Goals?

Energy markets work best when free market forces drive decisions on how oil and gas are produced, transported, and purchased.

This is normally the case for private firms and can even be the case for state-owned oil and gas companies. But governments can and should play a facilitating role. Governments should put in place the right business climate to attract investment and should work with neighboring states to expand the market and increase interconnectivities. We can encourage these efforts, and work with our partners to create the political framework in which businesses and commercial projects can thrive.

At the heart of our policy is the belief that energy security is best achieved through diversity – diversity of suppliers, diversity of transportation routes and diversity of consumers, together with a focus on alternative technologies, and renewable and other clean energy technologies, and increased energy efficiency.

THE SOUTHERN CORRIDOR

In line with this approach, the Obama Administration strongly supports the establishment of a new pathway, the so-called Southern Corridor, to bring natural gas to Europe, via Turkey, from the Caspian and potentially other sources beyond Europe's south-eastern frontiers. Gas from Azerbaijan's offshore fields will be the first significant volumes available to supply the Southern Corridor. Development of a major field is well underway by the international Shah Deniz consortium. Three separate pipeline consortia --the Nabucco, ITGI and TAP groups – are meanwhile laying the financial, technical and organizational groundwork to compete for the right to ship Shah Deniz gas. From the standpoint of US policy goals, the best outcome is one that brings the most gas, soonest and most reliably, to those parts of Europe that need it most. But at the end of the day, any solution for bringing Azeri or other gas to European customers must make commercial sense. In light of the momentum achieved over the past 18 months, we are confident that a commercially viable Southern Corridor will be realized. The investment decisions to make that possible should occur by the end of this year.

While gas from the Shah Deniz field will be the first gas available and is necessary to the development of the Southern Corridor, there is good reason to believe other sources can ultimately contribute additional volumes to the Southern Corridor. Azerbaijan has promising potential in other Caspian fields. Turkmenistan is investing in infrastructure to keep open a western gas export option. And there are increasing indications that, once it finds the means to satisfy pressing domestic demands for electricity, Iraq will have significant gas export capacity. We have an active dialogue with these countries, with key partners like Turkey, and with relevant members of the international business community, to explore and encourage development of the broadest possible basis for a robust Southern Corridor that will ultimately be a critical element in the global energy architecture.

Oil markets and oil production are also keys to our strategy. We actively support further production of oil in Kazakhstan and new export routes for that oil to world markets, since Kazakhstan is one of the few non-OPEC countries with the potential to substantially increase production over the next 10 years.

RUSSIA

Some people have portrayed our energy policy and Russia's as the next round in the Great Game in Central Asia. I reject this analogy. Energy security and energy investment are topics for serious two-way discussions with Russia. The importance of these issues is reflected by the inclusion of a Working Group on Energy, chaired by Energy Secretary Chu and Russian Energy Minister Shmatko, under our US-Russia Bilateral Presidential Commission. The Energy Working Group also includes sub-groups on Energy Security, Energy Efficiency, and Clean Energy Technologies.

Discussions held in the context of the Presidential Commission, and other venues, will help us find areas where the United States and Russia can mutually benefit from cooperation in the energy field.

Russian Deputy Minister of Energy Kudryashov and I chaired two meetings of the EWG's Energy Security Sub-Group, in which we have had productive discussion on issues such as world market energy trends, proposed projects to bring new oil and gas to market and bilateral investment in Russian and US energy markets. Russia is an important supplier of oil and gas, and we welcome market-driven expansion of its production capacity.

We are considering a Russian proposal to hold an energy investment conference in Moscow in the fall.

The Energy Efficiency Sub-Group, co-chaired by the Department of Energy (DOE), is developing a pilot smart grid project that will match Russian and US cities and facilitate the sharing of experiential data, technical information and best practices. The Clean Energy Sub-Group, also chaired by DOE, has focused on facilitating exchanges and sharing best practices with regard to biofuels, solar, new materials, smart grids, hydrogen economy, and clean energy forecasting. The Energy Efficiency and Clean Energy Sub-Groups have also agreed to cooperate on federal energy management programs and financing mechanisms, and to conduct joint research on clean energy technologies.

Ukraine

The United States and Ukraine are heavily engaged on energy issues. We discuss energy on a regular basis through the US-Ukraine Commission on Strategic Partnership as well as the US-Ukraine Energy Security Working Group. We continue to encourage Ukraine to take the necessary measures to attract foreign investors to its energy sector, including transparency, good governance and effective legal and regulatory frameworks. Ukraine has recently taken steps to reinforce its energy security, including plans to increase domestic supply of both conventional and unconventional gas sources. The February 2011 signing of the US-Ukraine Memorandum of Understanding on Unconventional Gas Resources marks a deepening of our bilateral cooperation in the energy sphere. The next step will be cooperation between the US Geological Survey and Ukraine on assessing Ukraine's unconventional gas potential, and providing advice on related environmental and regulatory issues

Central and Eastern Europe

We support Central and Eastern European states efforts to more effectively diversify energy sources. We are encouraging them to coordinate work toward a common energy market and to increase gas and electricity connections with each other and with the larger, better-supplied economies of Europe. To this end, the US is actively cooperating with the Baltic States in implementing the Baltic Energy Market Interconnection Plan (BEMIP), which strives for full integration of the Baltic States into the European energy market. The US also encourages increased regional efforts in the energy sphere by the Visegrad Four (V4); Poland, Czech Republic, Slovakia, Hungary, and other countries in the region. New gas import infrastructure is underway in Poland and elsewhere in the region.

Existing Infrastructures, EU Law and Unconventional Sources

It is important to emphasize that new pipelines alone will not full assure Europe’s energy security. The US supports the other initiatives that Europe is undertaking to increase its own energy security, including through a major set of decisions in energy that the European Council recently adopted. Those initiatives focus on building a single market for energy, unbundling the distribution and supply functions of energy firms, building interconnectivity of European gas and electricity networks, environmentally sound development of shale gas reserves, enhancing LNG import capabilities, increasing gas storage, improving energy efficiency, and exploring alternative and renewable sources. All of these are pieces of the puzzle to ensure European energy security.

While some of that appears highly technical, these measures move toward a future where every country in Europe can access regional and global markets for energy resources. We are already seeing this today, as the United Kingdom is becoming an important transit country for LNG supplies to reach continental Europe. Some of this included LNG that has become available to global markets as US domestic gas supplies have increased. Interconnector pipelines are being expanded or built to move gas across borders in multiple directions within Europe. Such developments should greatly improve Europe's ability to respond to supply shocks and to adapt to changes in global supplies and markets.

THE US - EU ENERGY COUNCIL

The US-EU Energy Council was launched in November 2009, formalizing our ongoing engagement with the EU and member states on energy issues in three areas: Global Energy Security and Markets, Energy Policy and Standards, and Technology and Research Cooperation. We have developed coordinated approaches to Ukraine, Russia, the Southern Corridor and Iraq. We are coordinating with the EU to encourage Ukraine to implement international financial institutions' pre-conditions for financing. The Energy Council played a key role in encouraging Turkey and Azerbaijan to reach an agreement on gas transit and supply conditions, which will ultimately help secure Caspian basin supplies via the Southern Corridor. We are also exploring options to foster development of Iraq's gas resources.

In coordination with the Trans-Atlantic Economic Council, the Energy Council is making strides toward harmonizing battery and plug-in charging standards for Electric-drive Vehicles (EVs) and software for Smart Grids. Finally, we are working to facilitate an unprecedented level of researcher exchanges in key areas of clean and renewable technologies research.

Given current global financial and economic conditions, it is more important than ever for countries to have reliable and consistent access to affordable energy supplies. Greater energy interconnectivity, competitiveness, and transparency will increase energy security and propel economic development and prosperity.

In sum, the key to achieving our Eurasian energy strategy is engagement. We need to continue to engage with the private sector, with the EU, and also with Turkey, Russia, and Caucasus and Central Asian countries. Our job is to listen, identify common interests and priorities and play a facilitating role where we can. We appreciate your interest in these matters and look forward to working with you to build stronger relationships with our partners around the world to ensure that we all enjoy an energy secure future.

Thank you.

In: European Energy Security Options ... ISBN: 978-1-62081-814-5
Editors: R. E. Phillips and J. B. Cook

Chapter 4

TESTIMONY OF ARIEL COHEN, SENIOR RESEARCH FELLOW FOR THE HERITAGE FOUNDATION. HEARING ON "EUROPEAN AND EURASIAN ENERGY: DEVELOPING CAPABILITIES FOR SECURITY AND PROSPERITY"*

My name is Ariel Cohen. I am Senior Research Fellow for Russian and Eurasian Studies and International Energy Policy at The Heritage Foundation. The views I express in this testimony are my own, and should not be construed as representing any official position of The Heritage Foundation.

As a giant energy producer and major energy transit country, Russia is an important player in the field of global energy production. Russia has the largest reserves of natural gas in the world, and possesses some 79.4 billion barrels of oil, approximately 6.4% of the world's total.[1] In 2009 oil production, Russia accounted for 9.9 million barrels per day (mbd), competing only with Saudi Arabia for the title of number one oil producer.[2] Total Russian net oil exports reached 7 mbd the same year.[3] Russia produced 527.5 billion cubic meters of natural gas during the recession in 2009, second only to the United States.[4] In addition, large areas of Eastern Siberia and the Arctic are still unexplored and, according to experts, are expected to yield up to a quarter of the world's energy supply.

---

* This is an edited, reformatted and augmented version of testimony Presented June 2, 2011 before the House Foreign Affairs Committee.

Despite its vast resource base and its formal assurances of its reliability as a partner, Moscow has already proved that it is willing to hike up oil and gas prices to match the general trend of higher energy prices, engage in anti-free market practices, especially at home and in Europe, and use energy as a foreign policy tool.

Russia is willing to use force to achieve its geo-economic goals as well. Control of energy corridors from the Caspian Sea to the Black Sea and beyond was an objective of the Russian military operation against Georgia in August 2008. This has been clearly confirmed by other incidents involving delays in energy supplies to Ukraine, Azerbaijan, Belarus, Georgia and the Baltic states. Many argue that Moscow‘s international energy behavior leaves its partners insecure and makes observers doubt that Russia is a responsible player, especially when unconstrained by competition and powerful investment sources.

To this day, Moscow is dependent upon the massive pipelines built during the Soviet era. Russia‘s energy policy is facilitated by the Soviet-era oil and gas infrastructure that ties Central Asian producers to Russia for their access to external markets. As part of its strategy, Russia works to maintain control over energy transportation routes and opposes any projects that could provide Europe with alternative supply lines. European demand was very high before the recent economic crisis, and is projected to grow further provided the current geopolitical instability does not cause another global recession. Eastern Europe consumes even higher percentages of Russian energy, with several countries entirely dependent on Russian gas. However, US and Europe should have serious reservations concerning Russian practices that limit access to the market, tend to promote corruption, and expand Moscow‘s energy transport agenda.

Russian Energy Policy Overview: Energy Nationalism and Abuse of State Ownership

Russia has criticized Europe‘s approach to international energy security as limited to the interests of energy importers. Under Russia‘s presidency in the Group of Eight (G8), then-President Vladimir Putin made energy relations a central theme at the 2006 summit in St. Petersburg, presenting his own vision for ―global energy security".[5] While talking of interdependence and dialogue,

Russia insisted on demand guarantees for the producers, and sharing responsibilities and risks among energy suppliers, consumers, and transit states. Putin spoke of joint commitments to work on the energy arena with coordination and distribution of profits and risks to prevent energy conflicts.[6] This would not be a problem if Russia allowed international oil companies (IOCs) minimally restricted access to its energy resources. Unfortunately, since 2003 this hardly has been the case, as the state has pursued a policy of monopolizing gas production and oil and gas pipeline transportation, and has tightened its grip on the quickly growing oil production sector by effectively expropriating YUKOS and buying the Sibneft and Russneft oil companies.

The trend toward marginalizing and even actively persecuting independent Russian energy businesses has continued, with the controversial re-sentencing of Russian oil magnate Mikhail Khodorkovsky in December of 2010, resulting in six more years in prison for the former businessman.[7] His appeal to the Moscow City Court was rejected on May 24th, 2011, despite the fact that President Medvedev publicly announced that Khodorkovsky does not represent a danger to public order. Moreover, after a seven-year delay, Amnesty International finally declared Khodorovsky a prisoner of conscience.[8]

The Khodorkovsky case is particularly symbolic. In 2003, the former oil tycoon went from being one of Russia's leading energy capitalists, owner of the YUKOS Oil Company and a promoter of economic and political liberalization, to a political prisoner. Of course, YUKOS was subsumed into Igor Sechin's Rosneft, a major state-owned oil company until recently. Before he was imprisoned, Khodorkovsky's opposition to then-President Vladimir Putin's authoritarian administration and the spread of corruption resulted in a long list of absurd charges of tax evasion and fraud. Somehow, even the fact that YUKOS was making a profit under his leadership became a chargeable offense.

The last conviction was handed down on December 29, 2010, after Khodorkovsky was accused of stealing 350 million tons of oil from his own company, a claim made all the more unbelievable given that the Russia state accounted for – and taxed – all the oil YUKOS sold, and that the total alleged production figures were higher than YUKOS' reported output.[9]

The tainted nature of this case became even more evident after lawyer and blogger Alexei Navalny's exposé of corruption in the construction of Transneft's massive East Siberian pipeline. Some four billion U.S. dollars have been stolen or defrauded by individuals close to the Kremlin with no redress.[10] Instead of thanking the whistle blower, the authorities paid Navalny back by launching a criminal investigation against him.

The non-transparent, unfair nature of the Khodorkovsky cases has received a great deal of criticism from Western leadership, including a statement from President Obama as well as European governments.[11] The matter highlights not only the —vendetta" politics of contemporary Russia,[12] but also that fact that the Russian government, not energy companies or international markets, sets policy on the nation‘s economy, and particularly its energy sector.

ENERGY SUPERPOWER?

Russia‘s Energy Strategy, adopted in 2003, sets the framework for the country‘s energy policy. Thus, Russian energy security builds upon —protection of the country, its citizens, and economy from [external and domestic] threats to the reliable energy supply," including geopolitical and energy market risk factors.[13] Moscow claims to promote a non-discriminative regime for the Russian companies to access foreign energy markets and advance their participation in large international oil and gas projects. Energy factors are placed in the center of Russian diplomacy. As then-President Putin noted in one of his speeches, —the place Russia takes in global energy cooperation directly impacts its current and future wellbeing."[14] Russia‘s energy strategy through 2030, formulated in 2009, predicted even greater increases in Russia‘s oil and gas production sector, despite issues with property rights, an impractical and unevenly enforced tax code, the huge scope of needed investments, and concerns about the obsolescent infrastructure.[15] With ample energy resources and a dominant position in the European market, Russia‘s hydrocarbon power will remain impressive into 2020 and beyond. Russia‘s decision makers sense that consumer governments and companies, anxious to gain access to coveted barrels and cubic meters, do not want to challenge the supplier‘s assertive foreign policy.

In the remaining years until 2020, Russia seeks to maximize its economic and geostrategic advantages as a major energy producer with vast hydrocarbon reserves. This becomes even more poignant as Middle East supplies are now suffering from the repercussions of the "Arab spring" and the future of the nuclear power has become more uncertain as a result of the nuclear power station disaster triggered by the recent tsunami that struck Japan.

The Kremlin has advanced Russia's energy strategy through an array of security and economic policies, all of which aim at a common strategic goal. The aggregate effect is to create customer country dependency by locking in demand with energy importers and consolidating oil and gas supplies by signing long-term contracts with Russian and Central Asian state-owned or state-controlled energy producers and Russian state-owned pipeline monopolists.

Europe is mainly concerned about potential supply disruptions resulting from government actions or policies that impact gas supply sources and transit. In recent years, the issue of gas corridor diversification has become increasingly important for Europe as EU officials try to reduce the region's dependence on Russian gas.

ARCTIC ENERGY STRATEGY

When the Russian flag was planted on the Arctic seabed under the North Pole in 2007, this was no pretense at subtlety. The act was overt and audacious. Moscow claims a sector of the energy-rich Arctic continental shelf along the Lomonosov Ridge, an underwater structure protruding from the northern coast of Eurasia towards the North Pole and abutting the Canadian and Danish sectors. Vladimir Putin weighed in during a speech on a Russian nuclear-powered icebreaker in early 2007, urging greater efforts to secure Russia's "strategic, economic, scientific and defense interests" in the Arctic.[16] Moscow's moves are dictated by energy-driven geopolitics and geo-economics. Geologists believe the Arctic Ocean's seabed may contain nearly 25 percent of the world's remaining hydrocarbon deposits. It is also rich in diamonds, and precious ferrous and non-ferrous metals.[17] As the ice cap melts and shrinks, these resources will become more accessible and a new sea passage along the northern coast of Eurasia may provide a cheaper transportation route.

From a geopolitical perspective, the exploration of polar petroleum reserves may be the kind of opportunity that allows Russia to become what then-President Putin termed "an energy superpower." Russia seeks to expand its continental shelf beyond the 200-mile economic zone through a mechanism provided by the UN Commission on the Limits of the Continental Shelf under the 1982 U.N. Law of the Sea Convention (UNCLOS), to which Russia is a party. Moscow claims that two underwater mountain ridges jutting into the Arctic Ocean from the Russian continental shelf—the Lomonosov Ridge and the Mendeleev Ridge—are extensions of the Eurasian landmass[18] . The span of this area of seabed approximates the size of France, Germany and Italy combined. Russia's first claim with the UN, submitted in 2001, failed due to insufficient evidence.

In recent years, Russia has aggressively moved forward with the expansion its presence in the Arctic region, while the US has been less active in advancing its interests in this strategically important region endowed with vast natural resources. [19] As Arctic sea-lanes are becoming more navigable due to climate change, the competition for the vast natural resources of the Arctic is more likely to intensify. In February 2011, Russia's state-controlled Rosneft and British petroleum giant BP entered into an agreement to develop Arctic oil fields with estimated reserves of 5 billion tons of oil and 10 trillion cubic meters of gas. However, Russian Arctic energy development is likely to face difficulties because of the significant risks and costs associated with Arctic offshore drilling.

Great challenges also abound for the Arctic and East Siberian/Far East gas fields. These reserves lack functioning gas fields and pipelines, and require hundreds of billions of dollars in investments. Many hopes were hung on the Shtokman gas field, located over 300 miles offshore in the Barents Sea, where local sea depths exceeding 300 meters.[20] After many delays, Gazprom reconsidered its earlier decision to develop the field alone. In July 2007, Gazprom signed an agreement with France's Total, followed by a deal with Norway's Statoil Hydro in October 2007 covering the first phase of Shtokman development.[21] , U.S. companies were kept out despite earlier promises to include Chevron and possibly Conoco Phillips. However, the agreement gives Total and Statoil Hydro no ownership rights to the gas. Gazprom, through its 100 percent-owned subsidiary Sevmorneftegaz, remains the full owner of the Shtokman development license and will be the full owner and sole exporter of products.[22]

While Norway's Statoil Hydro has vast experience drilling off shore in the northern longitude, Total is cash-rich but has no experience working in Arctic conditions.[23]The completion of the Shtokman field in the Arctic has now been pushed back to 2016. Other fields under development in the Arctic and polar regions are often even more challenging than Shtokman. Only in the case of the Kovykta field in East Siberia is production assured: this field was essentially expropriated away from BP by the GOR and handed over to Gazprom, which the GOR controls, so that Gazprom could develop it and build a pipeline to China. Likewise, there is substantial additional gas in

Eastern Siberia, including in Yakutia, which could be developed for the Chinese market.

China, Japan, and other destinations in East Asia are also attractive markets for East Siberian and Sakhalin Island gas, but it remains to be seen if Russia develops massive new fields in the Arctic, as the difficulties with the recent Rosneft-BP deal may suggest. There, BP‘s Russian billionaire partners in the TNK-BP joint venture derailed an asset swap and Arctic field development by BP and Rosneft.

Recoverable gas and oil reserves around Sakhalin Island, one of the world‘s largest natural gas fields, are estimated at almost 7 billion barrels and 80 trillion cubic feet respectively, one of the largest in the world.

The Russian government announced a number of costly programs to explore and develop the East Siberian oil and gas fields and to build a network of oil and gas pipelines in the 2020- 2030 timeframe, which will cost tens of billions of dollars.[24] In addition, the Russian leadership realizes the need to open up to foreign investment in its energy sector, since Russia needs Western capital and technology to successfully develop its climatically and geologically challenging oil and gas reserves. Furthermore, Russia, unlike any of the other major energy exporting countries, is also one of the world‘s leading industrial energy consumers, primarily because of the country‘s inefficient, aging infrastructure and utilities.

INTERNAL CONSOLIDATION

The Russian oil and gas sector is notorious for easing domestic and foreign corporations out of majority equity stakes in Russian mega-projects and for consolidating domestic ownership in the hands of government-controlled entities. The two Russian energy national champions— vertically integrated state-owned or controlled global companies capable of competing with foreign corporations—are headed by senior officials close to Vladimir Putin. Putin‘s former Chief of Staff and later successor as president, Dmitry Medvedev, was the ex-officio chairperson of Gazprom. Today, Putin‘s one-time mentor, former Prime Minister Victor Zubkov, occupies this position. Putin ally Alexei Miller is the long-serving CEO of Gazprom. Putin confidante and First Deputy Prime Minister Igor Sechin, who is in charge of energy and heavy industry, until recently chaired the board of Rosneft, Russia‘s largest state-run oil company, which expropriated the bulk of YUKOS assets. This management scheme ensures that Gazprom and Rosneft are reliable foreign policy arms for the Kremlin. Since the early 2000s Moscow limited access by major international energy corporations to giant Russian fields and forced them to give up their majority stakes in lucrative projects.

The Natural Resources Law limited foreign participation in energy exploration projects to minority stakes—25 percent in ‗strategic' oil and gas fields, and 49 percent in other energy projects. Limited in their rights to own exploration licenses, the transnational corporations are reduced in many cases to operator or technical service provider roles. In June 2007, then-First Deputy Prime Minister Sergey Ivanov said that foreign companies ―will never operate" Russia's major fields again.[25]

Although leading officials, including Mr. Medvedev, have explicitly rejected state capitalism as a model for Russia, the Kremlin is actively consolidating its ownership in the energy sector. Putin envisages the state not as the great re-nationalizer, but as the biggest shareholder in a privatized economy.[26] The return of strategic assets under state control is often presented to the public as the restoration of national property illicitly acquired in the mid-1990s by corrupt and politically manipulative oligarchs at deeply discounted prices.

This certainly was the case with state-owned Rosneft's 2004 murky acquisition of Yuganskneftegaz, the key production unit of forcibly bankrupted YUKOS. Despite the company's having received a clean bill of health from the Russian tax authorities, the State, through trumped-up bankruptcy proceedings, sold YUKOS to a straw company in a no-bid sale, which in turn sold it to Rosneft at a grossly undervalued price. Rosneft then amalgamated the YUKOS oil-producing company into its operations. The two principal YUKOS principal owners, Mikhail Khodorkovsky, discussed in detail earlier, and his business partner Platon Lebedev, received 13.5 year sentences in two highly flawed court proceedings; and the YUKOS affair became a byword for Russian-style judicial arbitrariness and politically motivated justice.

Western companies were also the subject of heavy-handed state interventionin the energy sector. Royal Dutch Shell was evicted from the Sakahlin project. British Petroleum was evicted from the lucrative Kovykta gas field in eastern Siberia after the forced sale of its 62.9 percent stake to Gazprom in June 2007. The TNK-BP joint venture was unable to meet production quotas prescribed by the Kremlin because pipeline monopolist Gazprom refused to develop any export pipelines. After officials threatened to cancel the license, and the courts refused to intervene, BP-TNK agreed to sell its Kovykta stake to Gazprom at a fraction of its market value.[27] Later on, in

2007–2008, the TNK-BP joint venture, with its unique 50–50 control between the Russian and British partners almost fell apart. This was due to pressure from the Russian partners, known as Alfa Access Renova (AAR) to oust the BP-appointed CEO and gain more control of the company. Many experts suspected that the ultimate goal was to force the British company to sell to AAR or to a Russian state-owned oil company; however, falling oil prices and the precipitous Russian stock market slide of 40 percent from May to August 2008 may have put pressure on the Russian partners to settle. A compromise, rare in the Russian oil sector, was achieved in early September 2008, and for now, the joint venture is continuing.

BP, however, signed what its chief executive, Bob Dudley, termed a —milestone" a joint-venture contract with Russia‘s Rosneft oil giant to develop the oil reserves in Russia‘s Arctic regions in January 2011.[28] However, relations between AAR and BP remain problematic, as the oligarchs sued BP in Great Britain and prevented its agreement with Rosneft from being executed. At this point, BP is trying to buy out AAR from the TNK-BP joint venture. While BP‘s leadership may envision a new strategy in Russia aimed at overcoming its troubles in the Gulf of Mexico, Russia‘s past relationship with foreign oil companies and with BP in particular raises many risk management questions.

Due to the resumption of global economic growth, and even before the current instability in the Middle East, oil prices have been increasing since January 2011, and consequently, Russian oil companies been enjoying higher oil revenues.

Sloviki in Charge

The Kremlin-affiliated structures are squeezing independent energy companies to get hold of their assets. In a ground-breaking interview to *Kommersant*, Oleg Shvartsman, then the head of the Finansgroup financial-industrial group close to the *siloviki* ("men of force", primarily the leaders around Putin who have security services backgrounds) revealed a scheme intended to pressure private companies that the Kremlin finds insufficiently accountable to the state.[29] Among the group‘s key assets is the Russian Oil Group that cooperates with Rosneft, TNK, and Lukoil. After an initial push for trading alliances, Finansgroup began to acquire small and medium-sized oil-refineries, using illicit activities to bring down corporate values prior to the acquisition.[30] Finansgroup also manages the so-called Social Investments

Corporation, exercising what Shvartsman called the "velvet re-privatization" of strategic assets based on various voluntary and coercive market instruments of asset absorption. Shvartsman said the group enjoys the full support of the Russian "power" ministries, including the Interior Ministry, FSB, and the tax and environmental authorities.[31]

Simultaneously, Russia is seeking to develop its energy services industry. Only recently, the Eurasian Drilling Company, the largest provider of onshore drilling services in Russia, signed a substantive contract with world-famous hydrocarbon oil and gas services firm Schlumberger. According to the strategic contract, a vast exchange of assets will occur between the two industry leaders, but most importantly, Schlumberger will become a subcontractor for Eurasia's drilling operations for "up to 200 rigs for a 5-year period."[32]

Domestic consolidation of the Russian oil and gas industry under the Kremlin's direct ownership or control increases Moscow's options in the continued use of energy as a foreign policy tool. These major takeovers and expropriations further limit opportunities for foreign investment and technology transfer to the Russian energy sector and beyond. They signal the return of statist economic policies, and widespread corruption, while allowing the state to interrupt the flow of oil or gas for political reasons far more easily than a private-sector owned company would be able to do.

RUSSIAN ENERGY GEOPOLITICS TO 2020: THE AMERICAN PERSPECTIVE

Russia's energy nationalism has been a source of frustration in Washington. From an American perspective, growing European energy imports on monopolistic Russian oil and gas exporters is a negative long-term geopolitical trend. However, there are other issues. Despite being the world's largest energy consumer, the United States has limited energy relations with Russia. In 2002–2003 Russia refused to construct projects dedicated to oil exports to the United States, such as the Murmansk pipeline, suggested by the then-privately held YUKOS, LUKoil and Sibneft oil companies. Moscow has also derailed attempts by U.S. oil supermajors to buy significant non-controlling stakes in large private Russian companies such as YUKOS. On the other hand, Gazprom considered and abandoned plans to export LNG to the US West Coast.

If Moscow is serious about the —reset|| , it needs to address a number of fundamental issues, first and foremost the issues of market access and the rule of law. First, US companies should gain access to oil and gas fields and pipeline projects, not limited by the obsolescent Natural Resources Law and the state-owned pipeline monopolies.

Second, for Russia to develop its Arctic and East Siberian reserves, it needs to put an end to corruption; completely revamp the rule of law, including assuring independence of the courts from the executive branch regardless of the scope of disputes and the powerful personas involved. The Russian government should provide Western companies enforceable guarantees that foreign investment will not be jeopardized by moving goalposts; revising the terms and conditions of prior investments. Ideally, it should also significantly disinvest the state from the natural resources sector. Finally, prior violations of the rule of law need to be redressed.

Energy issues spill over into the realm of the geopolitical balance-of-power. When energy prices skyrocketed in 2007-2008, Russia quickly evolved into an assertive anti-status quo power that challenged the U.S. and its allies on many fronts, especially in the territory of the former Soviet Union, as the 2008 Russian-Georgian Five Day War demonstrated. There are also ongoing frictions in the Balkans and the Middle East, where Russia has opposed Western policies. This happens both because of the ample funding available to finance a more ambitious foreign policy due to energy revenues and the self-assurance which comes with general economic prosperity, as well as from Moscow‘s tendency to use energy as a foreign policy tool. As oil prices rise, it is safe to expect Russia‘s cockiness to return.

Russia‘s strategic goals include preventing countries around its borders from becoming pro-American as well as increasing control over the transportation of Russia hydrocarbons through the territory of its neighbors. Furthermore, the Kremlin aims to control the export of oil and gas from neighboring countries by directing their flow via the Russian pipeline system. By locating pipelines and gas storage facilities in Austria, Hungary, Bulgaria, Greece and Turkey, Russia connects them to Moscow with ties that bind. Sometimes, these ties also include lucrative personal economic deals, as demonstrated by the employment of Gerhardt Schroeder as Chairman of the North Stream gas pipeline consortium, and similar arrangements for other prominent European politicians.

Russia also attempted to push the United States out of Central Asia, and successfully limited US participation in new Caspian energy projects, excluding it from the SCO‘s Energy Club. The United States, for its part, supports diversification of energy transportation routes in Eurasia. From the Russian perspective, the U.S. and EU-backed pursuit of diversified energy sources and transportation routes is unfriendly towards Russia, politically motivated, economically unfeasible and environmentally damaging. The Kremlin is likely to use Europe‘s dependence on Eurasian energy to exacerbate differences in transatlantic relations and use its influence to minimize the pro-American foreign policy agenda. In the current decade, America‘s allies in Europe may face tough choices between the cost and stability of their energy supply, on one hand, and siding with the United States on key issues, on the other hand.

In sum, the developed world economies and energy net importers in general will benefit from greater stability, security, transparency, and the rule of law in energy-exporting states, to ensure that oil and gas remain readily available, ample, affordable, and safe. However, the Kremlin views energy as a tool of assertive foreign policy and uses it broadly, often without much concern for diplomatic niceties. If current trends prevail, this decade may see the Kremlin might translating energy monopoly into increasing foreign and security policy influence in Europe. In particular, Russia is seeking recognition of its predominant role in the post-Soviet space and Eastern Europe. This has already affected geopolitical issues important to the West, such as NATO expansion, ballistic missile defense, the tension around the status of Kosovo, and Moscow‘s increasing influence in the post-Soviet space.

Furthermore, before the current instability erupted, Moscow was seeking to re-engage in a centuries-old balance-of-power game in the Middle East, from Algeria, where it attempted a gas condominium, to Syria, where it is rebuilding naval bases in Tartus and Ladakiye and supplying modern weapons, to Iran and India.[33] Though in the end it voted in the UN Security Council with the U.S. and Europe during diplomatic crises over the Iranian nuclear enrichment program, Moscow has continually provided Teheran ample diplomatic cover in the United Nations and elsewhere, as well as expanded arms supplies. Moscow also abstained in the UNSC vote on Libya – together with China and Germany. Premier Putin harshly criticized the Libya war as a ―new crusade", clearly trying to score points for Russia in the Middle East.

The Obama Administration seems to be less concerned about European energy dependence on Russia than its predecessor was. Just recently, Russian President Dmitry Medvedev and the American president announced plans to liberalize visa restrictions for businessmen and tourists travelling between the two countries in a joint statement at the G8 summit in Deauville. "We have instructed our officials to concentrate on visa liberalization on a reciprocal basis for the largest segments of our traveling nationals," the statement said. Under the new agreement, eligible business travelers and tourists would be issued visas valid for 36 months at a unified and reciprocal fee. Government officials would also be eligible to receive 12-month multi-entry visas.[34] While freedom of travel is a good thing overall, the US Congress is seeking to limit travel to the US on the part of clearly corrupt Russian officials or those engaged in human rights violations or property expropriation, as will be discussed *infra*. In this respect, the Administration is defying Congress and going in a wrong direction.

There are a number of steps the Obama Administration and Congress can take to address Russian state graft and corruption. Some of these measures were outlined by The Heritage Foundation.[35] Specifically, the Administration should increase cooperation among U.S. and allied intelligence services, law enforcement agencies, and independent experts to track Russian state and oligarch money laundering activities, corruption, and unfair competition practices. The White House should encourage U.S. and other multinational companies to compete in economically viable energy and infrastructure projects overseas through free-trade, diplomatic and security support, and regulatory and tax policies that will enhance free competition without government-directed investment decisions. U.S. companies should be encouraged to compete for participation in the development of Turkmenistan's gas fields as well as in other geopolitically significant ventures, which Russia is targeting in India, Southeast Asia, Africa, and Latin America. The US should promote market-viable alternative energy sources and unconventional sources of fuels worldwide to counter the West's strategic dependency on Russian, Iranian, and Venezuelan oil. This should be accomplished through deregulation and trade and tax policies that encourage innovation and investment to develop, and through commercializing new sources of energy that best meet the needs of individual regions and nations. Western economies will be better off by expanding the supply of transportation fuels and reducing their Russian energy imports, thus reducing the influx of revenue into Kremlin coffers. Last but not least, the US should expand security cooperation with Russia's energy-exporting neighbors and other countries that Russia is targeting for energy cooperation, including train-and-equip programs for military and security forces protecting pipelines, and officer corps education in U.S. military colleges, and should make better use of NATO's Partnership for Peace program.

Washington should encourage Europe's energy diversification, providing political and diplomatic support to major consumers of Russian oil and gas to develop alternative energy pipelines throughout Europe and Eurasia. It is vital for EU members to come up with a joint position on energy geopolitics instead of lucrative bilateral deals, which only increase the region's dependence on Russian oil and gas. It is also necessary to insist that Russia live up to its commitments to uphold and implement the rule of law, without which its economic development, property rights, and civil liberties will remain in limbo. Otherwise, Russia will be positioned to apply the ancient Roman principle—*divide et impera*—to 21st century energy geopolitics.Finally, Congress should pass – and the president should sign – S. 1039, the Sergei Magnitsky Rule of Law Accountability Act of 2011. This bill, dedicated to the memory of a Moscow whistleblower lawyer who died in jail under torture in November of 2009, revokes visas and prohibits financial transactions for Russian officials who engage in extrajudicial killings, torture, or other gross violations of human rights committed against individuals seeking ... to expose illegal activity carried out by officials of the Government of the Russian Federation; or ... to obtain, exercise, defend, or promote internationally recognized human rights and freedoms, such as the freedoms of religion, expression, association, and assembly and the rights to a fair trial and democratic elections (Section 4, Paragraph 2).

Congress has an important role to play in changing the relations with Russia in the energy field for the better, for the benefit of the Russian and American peoples and our European allies. It is time to make the Russian oil and gas sector more transparent and open to foreign investment while curbing the use of energy as a geopolitical tool, which endangers Russia's neighbors.

# # #

-- Ariel Cohen, Ph.D., is Senior Research Fellow in Russian and Eurasian Studies and International Energy Policy at The Heritage Foundation (www.heritage.org). He is a member of the Council on Foreign Relations and the author of 5 books and monographs, 29 book chapters, and over 500 articles on topics ranging from geopolitics and energy in Eurasia and the Middle East, to the rule of law and terrorism.

****

The Heritage Foundation is a public policy, research, and educational organization recognized as exempt under section 501(c)(3) of the Internal Revenue Code. It is privately supported and receives no funds from any government at any level, nor does it perform any government or other contract work.

The Heritage Foundation is the most broadly supported think tank in the United States. During 2010, it had 710,000 individual, foundation, and corporate supporters representing every state in the U.S. Its 2010 income came from the following sources:

Individuals	78%
Foundations	17%
Corporations	5%

The top five corporate givers provided The Heritage Foundation with 2% of its 2010 income. The Heritage Foundation's books are audited annually by the national accounting firm of McGladrey & Pullen. A list of major donors is available from The Heritage Foundation upon request.

Members of The Heritage Foundation staff testify as individuals discussing their own independent research. The views expressed are their own and do not reflect an institutional position for The Heritage Foundation or its board of trustees.

End Notes

[1] BP 2007 Energy Survey, June 2008, p. 6, http://www.bp.com/liveassets/bp_internet/ globalbp/ globalbp_uk_english/reports_and_publications/statistical_energ y_review_2008/STAGING/local_assets/downloads/pdf/statistical_ review_of_world_energy_full_ review_2008.pdf

[2] EIA, *Russia Gas*, Country Analysis Briefs, Department of Energy, Energy Information Agency, November 2010, http://www.eia.doe.gov/emeu/cabs/Russia/Oil exports.html (Accessed February 28, 2011)

[3] *Ibid.*

[4] BP 2009 Energy Survey, June 2010, p. 22, http://www.bp.com/liveassets/bp_internet/globalbp/ globalbp_uk_english/reports_and_publications/statistical_energ y_review_2008/STAGING/local_assets/2010_downloads/statistical_review_of_world_ener gy_full_report_2010.pdf

[5] Nina Kulikova, —*Voprosy Energeticheskoy bezopasnosti – pozitsiya Rossii* [The Issues of Energy Security – Russia's Position]," *RIA Novosti*, September 1, 2006, http://www. rian.ru/analytics/20060901/53406077.html

[6] *Global Energy Security Fact Sheet*; Official Website of G8 Summit in St. Petersburg, 2006, http://www.en.g8russia.ru/press/facts/global energy/ (Accessed March 5, 2011)

[7] Vasilyeva, Nataliya, Russian Tycoon Khodorkovsky gets 6 more years, *Washington Times*, World News section, December 30, 2010, http://www.washingtontimes.com/news/ 2010/dec/30/russian-tycoon-khodorkovsky-gets-6- more-years/ (Accessed March 1, 2011)

[8] —Amnesty International declares Khodorkovsky 'prisoner of conscience'", *RIA Novosti*, http://en.rian.ru/russia/20110524/164210044.html (May 28, 2011)

[9] Simon Shuster; Khodorkovsky Case: Russia's Courtroom of the Absurd; *Time*, Dec. 27, 2010, http://www.time.com/time/world/article/0,8599,2039824,00.html

[10] Volchok, Dmitry; Юрист Алексей НаВальный – о коррупции В компании "Транснефть" [Jurist Aleksei Navalny – on corruption in Transneft]; *Рааио Сеобоаа [Radio Liberty, Russian language]*, November 20, 2010, http://www.svobodanews.ru/content/ article/ 2225552.html

[11] Ariel Cohen; —Free Khodorkovsky"; *Heritage Commentary,March 10, 2010, http://www.heritage.org/research/commentary/2010/03/free-khodorkovsky*

[12] Tisdall, Simon. —Mikhail Khodorkovsky: The latest victim of Vladimir Putin's vendetta politics"; *The Guardian*, December 27, 2010, http://www.guardian.co.uk/world/ 2010/dec/27/mikhail-khodorkovsky-vladimir-putin

[13] —Energy Strategy of the Russian Federation for the Period till 2020," http://www.minprom. gov.ru/docs/strateg/1.

[14] V. Salgin, —Globalnaya energeticheskaya bezopasnost' I vneshnyaya energeticheskaya politika Rossii [Global Energy Security and Russia's Foreign Energy Policy],", *Neftegaz,* June 28, 2007, http://www.neftegaz.ru/analit/reviews.php?id=548

[15] Blagov, Sergei, Reality Check for Russian Oil, *Asia Times Online*, Central Asia section, January 5, 2011, http://www.atimes.com/atimes/Central Asia/MA05Ag01.html (Accessed March 1, 2011)

[16] Ariel Cohen, "Russia's Race for the Arctic," Heritage Foundation *WebMemo* No. 1582, August 6, 2007, http://www.heritage.org/Research/RussiaandEurasia/upload/wm 1582.pdf

[17] Alexander Gabuev, "Print – Cold War Goes North: Russia and the West begin the race for the Arctic Region," *Kommersant,* August 4, 2007

[18] —Russia: Polar Expedition Means _Very Little' For Territorial Claims," RFE/RL, August 3, 2007

[19] Ariel Cohen. —From Russian Competition to Natural Resources Access: Recasting U.S. Arctic Policy," Heritage Foundation, Backgrounder #2421, June 15, 2010, at http://www. heritage.org/Research/Reports/2010/06/FromRussian-Competition-to-Natural-Resources-Access-Recasting-US-Arctic-Policy# ftn62 (March 16, 2011)

[20] Shtokman Field, Gazprom Project – Field website, http://www.gazprom.com/ production /projects/deposits/shp/

[21] —Statoil Hydro to develop Shtokman field," *New Europe*, Issue 753, October 25 2007, http://www.neurope.eu/articles/Statoil-Hydro-to-develop-Shtokman-field-/79155.php, and —Total signs on Shtokman dotted line," Upstream Online, July 13, 2007, http://www. upstreamonline. com/incoming/article137232.ece

[22] —New owner of Shtokman license", *Barents Observer*, Energy, February 18, 2009, http://www. barentsobserver.com/new-owner-of-shtokman-license.4559032-16178.html

[23] —Shtokman Watchers: Statoil Demonstrates Arctic Capability", *Oil & Gas Eurasia*, July 25, 2007, http://www.oilandgaseurasia.com/news/p/0/news/835

[24]U.S. Energy Information Administration, Independent Statistics and Analysis, Russia: Country Analysis Brief, November 11, 2010, accessed at: < http://www.eia.doe. gov/cabs/Sakhalin /Background.html>.

[25] Torrey Clark and Lucian Kim, —Gazprom Gains BP Gas Field as Putin tightens control," *Bloomberg*, June 22, 2007, http://www.bloomberg.com/apps/news?pid=newsarchive &sid=afw FHAGto f3Y &refer=uk

[26] Nick Paton Walsh, —Meet the Chief Exec of Kremlin Inc.," *The Guardian*, July 6, 2005

[27] —Russian Arm Twisting: Another Energy Firm Backs Down," *The Economist, June 22, 2007.*

[28] ABC Staff; *BP signs deal with Russian oil giant*; Australian Broadcasting Corporation, News, Jan 15, 2011, http://www.abc.net.au/news/stories/2011/01/15/3113634.htm (Accessed March 1, 2011)

[29] "*Partiyu dlya nas olitsetvoryaet silovoy blok, kotoryy vozglavlyaet Igor Ivanovich Sechin* [The Party is embodied for us in the power bloc led by Igor I. Sechin]." *Kommersant*, November 30, 2007, http://www.kommersant.ru/daily.aspx?date=20071130,

[30] Jonas Bernstein, "Finansgroup: How Russia's Siloviki Do Business," Jameston Foundation *Eurasia Daily Monitor,* November 30, 2007, www.jamestown.org.

[31] *Partiyu dlya nas olitsetvoryaet silovoy blok, kotoryy vozglavlyaet Igor Ivanovich Sechin* [The Party is embodied for us in the power bloc led by Igor I. Sechin]." *Kommersant*, November 30, 2007, http://www.kommersant.ru/daily.aspx?date=20071130

[32] *Eurasia Drilling Company and Schlumberger Enter Strategic Alliance, Sell and Purchase Assets*, Schlumberger, Press Release, October 5, 2010, http://www.slb.com/news/press releases/2010/2010 1005 edc slb alliance.aspx (Accessed on March 1, 2011)

[33] —Top Russian arms clients: India, Algeria, China, Venezuela, Syria", *World Tribune*, News, January 29, 2010, http://www.worldtribune.com/worldtribune/ WTARC/2010/eu_ Russia 0069_01_29.asp.

[34] —Medvedev, Obama vow to liberalize bilateral visas", http://en.rian.ru/russia/ 20110526 /164249981.html

[35] Ariel Cohen, Ph.D. *and* Lajos Szaszdi, Ph.D. —Russia's Drive for Global Economic Power: A Challenge for the Obama Administration", Backgrounder #2235, *January 30, 2009* http://www.heritage.org/Research/Reports/2009/01/Russias-Drive-for-Global-Economic-Power-A-Challenge-forthe-Obama-Administration

In: European Energy Security Options ... ISBN: 978-1-62081-814-5
Editors: R. E. Phillips and J. B. Cook

Chapter 5

TESTIMONY OF KEITH C. SMITH, CENTER FOR STRATEGIC AND INTERNATIONAL STUDIES. HEARING ON "EUROPEAN AND EURASIAN ENERGY: DEVELOPING CAPABILITIES FOR SECURITY AND PROSPERITY"*

SUMMARY

A major challenge to the new democracies of Central Europe is the corruption and lack of transparency in importing oil and natural gas from Russia and other energy producing states once part of the Soviet Union. This situation threatens to undermine good governance and ethical business practices throughout the European Union. It is a mistake to believe that there will be a major shift in the near term toward more openness in Russian foreign economic and political relations. Change will have to come through stricter EU enforcement policies and the adoption of stronger anti-corruption measures within the importing states. Unconventional gas production in Europe can help provide greater energy security in the medium term, but will not by itself reduce corruption or bring greater transparency.

EU membership provides only limited energy security to the EU's newest member states. The European Union lacks a common energy

* This is an edited, reformatted and augmented version of testimony Presented June 2, 2011 before the House Foreign Affairs Committee.

market and needs strong enforceable transparency and competition policies regarding energy imports. Nor does it have a common energy strategy concerning accountability by large importers, such as the Russian state monopolies Gazprom and Transneft.

Wealth accumulation from the energy trade is often used by powerful groups in the East to buy support in Western countries for Russian economic and security policies. This situation is facilitated by the lack of legal reporting requirements in Europe, including within the EU, concerning outside funding of political and business groups. Major Russian energy firms are busy financing large public relations efforts, university research and "independent" think tanks in Europe. This is one area where Europe could adopt laws to increase transparency, such as the US Foreign Agents Registration Act.

The most serious threats come from the danger of intervention at any point in the commercial process on the part of elite cartels that dominate the energy trade, particularly in Russia, Ukraine, and Central Asia. These cartels are composed of governmental leaders, intelligence officials, and favored business oligarchs. Lawsuits by these elites designed to silence Western critics have been a standard tactic up to now, including some oligarchs with close ties to Russian organized crime.

Courageous steps are needed by EU and Central European political leaders in order to bring about greater transparency and competition. Reformers too often pay for making the tough decisions by being turned out at the next election. After adopting badly needed improvements in legal and judicial systems they face well-funded opponents who are backed by nebulous firms representing energy importers. Real reformers would benefit from increased business transparency and from faster implementation of EU law concerning monopolies and accounting standards. The energy security concerns of the newer and more vulnerable member states have taken a back seat to the interests of the wealthier, but more energy secure countries in Western Europe.

CORRUPTION AND NON-TRANSPARENCY

Western energy firms too often confront demands from governmental officials and Eastern energy companies to engage in shady business practices when considering investment decisions, long-term sales contracts, and accounting procedures. Lack of a common energy policy and of strong antimonopoly enforcement regarding energy importers, have over the past 20 years increased opportunities for non-transparent state companies to secure influence among Western governments and with political and economic elites in Europe.

Dubious or outright corrupt business practices are distorting the energy decision making processes in both consumer and supplier countries. According to Russian economists, the business climate within the Russian energy sector has become less transparent and more corrupt in recent years. Why should one assume that this increase in Russian corruption has not spilled over into Europe, particularly into countries heavily dependent on oil and gas imports from, or through, Russia?

Many businessmen in former Communist states engage in what people in the West refer to as corrupt practices as a result of having grown up in a culture that considers bribery and coercion to be normal methods of conducting business. Some other influential individuals have close economic and personal ties that go back to the Soviet period. In addition, Western leaders are often at a significant disadvantage in negotiations with a highly trained former Soviet or Warsaw Pact intelligence officer. It is worth noting that both Rosneft and Transneft have hired "sleeper agents" recently expelled from the US for spying.

Elite Beneficiaries

Corruption and tight Kremlin control of Russia's energy companies have weakened the bargaining position of Western firms that normally use "best practices" when engaging in East-West energy trading. The domestic monopoly power and designated export privileges of Russia's state-controlled energy companies have led to a marked reduction of alternative investment possibilities for Western companies. This further intensifies the pressure on Western firms to agree to demands by Eastern energy suppliers that they engage in behavior not acceptable when dealing with other Western firms. In addition, this fuels the temptation for Western governments to ignore questionable business practices by their own domestic energy firms engaged in trade with Russian suppliers.

Western firms are already at a disadvantage when making business decisions in the former Soviet area. The absence of a win-win business concept in the East, combined with the absence of impartial judicial systems to enforce internationally recognized contracts between business firms, gives nontransparent firms an advantage in negotiations. Western businesses are usually compelled to work with partners favored by elite cartels. The existing systems appear to be designed to enrich networks of higher-level elites and/or their political parties, making it even more difficult for reform elements in some EU member states to bring about political and economic change.

WEAK WESTERN REACTION

The ineffective reaction of Western governments and the European Union to non-transparent business practices by Moscow only encourages the Kremlin to believe in the effectiveness of its aggressive energy policies and in its own ability to circumvent the "unbundling" and "third party access" rules of the EU.

The EU's blessing of the Nord Stream gas pipeline and what now appears to be its acceptance of South Stream, only facilitates continuation of monopoly and antitrust practices on the part of Russian companies. These antitrust and anti-competition practices are a clear violation of the original EC Treaty, the Energy Charter Treaty and now of the Lisbon Treaty. The EU's granting of "opt-outs" from its own rules only further encourages Russian firms to demand exemption from EU anti-monopoly rules.

Western governments continue to be reluctant to investigate and enforce the anti-bribery laws and regulations of the European Union and the Organization for Economic Co-operation and Development (OECD). The lack of a common EU approach to Russian and Central European energy policies allows Moscow to carry out a "divide-and-conquer" strategy that plays to the particular vulnerabilities of each European state.

Most EU member countries do not require their national firms that accept payments from foreign governments or that companies report this information to official government agencies and make the details concerning such payments available to the public. This makes it virtually impossible for governments, the media, or private researchers to know who is behind foreign-directed public relations campaigns or political influence peddling.

The problem is particularly pronounced in countries where the media is either unable or unwilling to engage in aggressive investigative reporting. Many newspapers and television stations in the new democracies are in precarious financial positions and are vulnerable to offers of help from well-financed foreign companies or their intelligence services.

The EU organizations in Brussels are particular targets of Eastern intelligence services that in many cases are charged with promoting the interests of their country's energy companies. Lobbying and public relations firms in Brussels are frequently hired directly and indirectly to further the interests of Gazprom and other Russian companies. Russian nationals are regularly employed by the European Union, some of whom then go on to lobby EU bodies on behalf of their country's energy firms.

Competition and Transparency

The new democratic states of Central and Southeastern Europe have too often been relatively passive in dealing with transparency and anticorruption issues. One reason for this could be that there are large numbers of political and economic leaders in the region who are holdovers from the Communist period. In the chaos resulting from the collapse of the Soviet Union and the Warsaw Pact, many former officials were able to gain control of the Communist Parties' assets and/or purchase large domestic industries cheaply. Not surprisingly, the business ethics of most former Communist officials did not significantly change with the fall of the old system.

Of course, the leaders of the new democracies are already faced with many problems, not the least of which is implementing the multitude of laws and regulations required for EU membership. Reformers find it difficult to curb the activities of the old-new business elite, particularly when these people already possess much of the countries' wealth and influence and often control large parts of the media.

Murky Alliances

The weak state of transparency in Central Europe aids the formation of new alliances between East European elites and former Communist Party and intelligence officers in Russia who dominate the major energy companies. This again puts Western firms at a clear disadvantage when negotiating for facilities acquisition or pipeline construction. With the renationalization of Russian energy assets, negotiations with the West are often carried out by top Kremlin officials. Fewer agreements are carefully negotiated at the company level. It is fair to ask whether there are many Western political leaders who can negotiate effectively with the seasoned intelligence officers in the Kremlin who determine Russia's energy policies.

Western leaders rarely have the skills or the ability to fully mobilize the state's resources in order to negotiate on an equal basis, and they often lack good intelligence regarding the tactics of the other side. Russia's top leadership is directly involved in energy sector deals with other countries. Prime Minister Putin devotes more time and attention to promoting his country's energy interests than does any leader from a Western country. Putin knows how to effectively use the Russian intelligence services to promote pipeline projects and downstream acquisitions.

In sum, corruption in the East-West energy trade will only end or diminish significantly when the large import countries decide that it is in their collective interests to abandon the practice of looking past non-transparent practices. They must be persuaded to stop attempting to give their own countries' energy firms a competitive edge at any cost. If they demand "best practices" by their own commercial interests and closely monitor the activities of government officials engaged in energy transactions, the more vulnerable "new democracies" will be more inclined to follow. When the large and more prosperous countries of Europe ignore corruption in the energy trade, they indirectly facilitate coercion of the new and smaller states of the region by more nontransparent, state-directed energy importers and foreign investors.

Saved by Unconventional Gas?

No doubt, the unconventional gas "revolution" that started in the US is already having an effect on the natural gas market in Europe. The US no longer needs LNG imports. This frees up LNG supplies to compete with piped gas from Russia. Prospects are good that significant quantities of shale and methane gas will be produced in Poland and the UK. Gazprom, however, is actively attempting to dampen enthusiasm for this new technology, and the company's export head, Alexander Medvedev, has warned Europeans of the alleged environmental dangers of unconventional gas production. Of course, Gazprom's leverage will be weakened by larger domestic gas production in Europe. However, the nuclear disaster in Japan will likely slow or stop the construction of new nuclear plants, thereby preserving the market for Gazprom exports to Europe over the medium term.

Environmental opposition to halt unconventional gas exploration, however, has sprung up in France and Germany. In many cases, it is difficult to trace the funding sources of these groups. A stop has been imposed on "fracking" efforts in those two countries, and a moratorium may be imposed on other countries, possibly by the EU's own environmental directorate. An official at DG Environment has even claimed that unconventional gas is "dirtier" than Arctic gas that will go through Nord Stream. This could be an attempt to dampen enthusiasm for increased exploration efforts within the EU. Europeans and Americans should carefully study the relative risks posed by the various forms of energy production. Unfortunately, too often frightening media headlines and poor science are utilized by various groups for their own political agenda. These scare tactics are encouraged by domestic coal producers and by foreign business interests, including Alexander Medvedev, the Director of Gazprom Export.

The views expressed above are solely those of the author and do not necessarily represent those of the Center for Strategic and International Studies.

In: European Energy Security Options ... ISBN: 978-1-62081-814-5
Editors: R. E. Phillips and J. B. Cook

Chapter 6

TESTIMONY OF AMBASSADOR ROSS WILSON, ATLANTIC COUNCIL OF THE UNITED STATES. HEARING ON "EUROPEAN AND EURASIAN ENERGY: DEVELOPING CAPABILITIES FOR SECURITY AND PROSPERITY"*

Mr. Chairman, thank you for this opportunity to be part of today's discussion about energy, Europe and Eurasia. Eurasia is part of a vast region of instabilities whose evolution will have a lot to do with whether the 21st century is reasonably peaceful or not, and energy issues are crucial for the global economy. It is wise that you focus on how these items come together and what should be American strategy going forward.

For nearly twenty years, the United States has pursued a Eurasian energy strategy based on four sets of interests and objectives.

First, given the dependence of our economy and the economies of most of our allies on imported energy, we have attached importance to the development of new sources of oil and gas. The political changes that came after the Cold War made available to world markets Caspian and Central Asian resources that were previously inaccessible. Beginning with talks that Secretary of State James Baker and US Ambassador to the Soviet Union Robert Strauss had in Almaty with Kazakhstani leader Nursultan Nazarbayev

* This is an edited, reformatted and augmented version of testimony Presented June 2, 2011 before the House Foreign Affairs Committee.

even before the collapse of the USSR, American officials and business leaders have engaged intensively to support Eurasian energy development.

Second, we have regarded the success of the new independent states of the former Soviet Union to be in America's interests. For over four decades, the principal threat to our way of life came from a Eurasian landmass dominated by a hostile Kremlin. More recently, new threats came from al-Qaeda, a network based in Afghanistan, but with offshoots in Central Asia. These post-Soviet countries' ability to govern effectively, secure their borders and provide better lives for their citizens matters to US national security. Energy is a potential "get out of poverty" card – an asset that resource and transit countries can use on behalf of a better future.

Third, we have supported multiple pipelines. This was summed up in bumper stickers an American ambassador to Kazakhstan, Beth Jones, had printed up years ago that read, "Happiness is Multiple Pipelines." Our policy recognized that economies depending on a single set of pipelines that go through one country, Russia, will be neither strong nor independent, and their relationships with Russia won't be healthy either. To express this positively, multiple routes, especially those like Baku-Tbilisi-Ceyhan (BTC) that lead directly to international markets, are the deliverers of real economic independence. They represent greater energy security for our European allies, too, and multiple pipelines are themselves spurs for further exploration and development.

A fourth set of US interests in Caspian energy could be ungrammatically summed up as "not Iran." It has long been US policy, reflected in legislation that this committee has been associated with, to oppose the development of Iran's oil and gas industries. Caspian energy, especially gas, has been a positive complement to the threat of sanctions – an alternative to Iran for US friends and allies desperate to augment and diversify their energy supplies. By helping develop pipelines to the Black and Mediterranean Seas, we have ensured that Iran could not become a principal export route for Caspian energy. Energy development and effective market-democratic policies in Eurasia have been and are ways to counter Iranian influence in Eurasia.

The next phase of Caspian energy development is natural gas, and it needs American support. The operator of Azerbaijan's offshore Shah Deniz project expects to decide in coming months on fully developing this giant gas field and on an export route to Europe for its gas. Negotiations among producers, potential consumers and transit countries on the details – and bickering over how to divide the spoils – have made progress slow. Patient US diplomacy led by Ambassador Richard Morningstar has proven very helpful.

The realization of a major new Southern Corridor gas export route and the successful exploitation of Shah Deniz as a supplier to Europe will significantly advance American interests. Among other things, this progress opens the door to further progress on Eurasian energy, including to connect Turkmen and northern Iraqi gas with European and other markets.

However, there is a serious possibility that well-intentioned new Iran sanctions legislation now under discussion may imperil this future. We risk undermining, if not standing on its head, a remarkably successful energy strategy that has helped our friends and allies and hurt Iran. Surely this is unintentional.

In the mid-1990s, Azerbaijan's president and a friend of the United States, Heydar Aliyev, decided on Western firms to lead the development of his country's large, offshore oil deposits at Azeri-Chirag-Guneshli (ACG), the BTC main oil pipeline, and Shah Deniz and its associated gas pipeline infrastructure. He made sure the planned pipeline routes would avoid Russia and Iran. He also thought it politic to give modest and non-controlling shares in ACG and Shah Deniz to Russia and Iran. One can argue from the vantage point of hindsight about whether this was wise policy. At the time, Azerbaijan felt its security interests could be directly threatened by both and so needed to placate them at a time when no other option for doing so was available. As I understand it, Iran sanctions language now under discussion might require BP and other Western firms with a stake in Shah Deniz to divest themselves of their investments and walk away from the project.

What will be the effect of this? Shah Deniz might not get developed. That will deprive our European friends and allies of needed gas and increase pressure on them to turn to Russia and Iran for alternative supplies. Or Azerbaijan might go ahead with the project, but to make money it would likely have to export the gas to Russia or to Iran – unhelpful for US policy on Eurasian energy or for punishing Iran. Without Shah Deniz, other Caspian gas will not get developed. No new, larger volume gas pipelines from the Caspian to Europe will be built. No Turkmen gas will flow west, little or no gas from Iraq will flow north, and the energy security of America's allies in Europe will remain more uncertain than is wise or necessary. US relations with Azerbaijan, a close partner on regional energy issues for nearly two decades, will suffer, with ill effects for that country's cooperation with us on Northern Distribution Network supplies for Coalition forces in Afghanistan.

I urge Congress and the Administration to work with interested parties on Iranian sanctions legislation that will be strong and effective, to be sure, but that also will not sacrifice important and longstanding US objectives and interests with respect to Shah Deniz and Eurasian energy.

Mr. Chairman, recent events in the Middle East and North Africa remind us that no energy strategy in any part of the world can ignore domestic politics or economics. In the 1990s, I and others argued that these new countries were on democratizing and market liberalizing trend lines, even if on many specific issues there were serious failings. The United States had a rich and frequent conversation at senior levels with leaders from all across Eurasia. There was buy-in on issues of interest to them *and* on issues of interest to us, including on economic and political liberalization. Virtually nowhere can one say these things now.

To ensure a reasonable level of success in these still fragile and ill-developed countries in a region marked by instabilities, we need more intense, sustained and broadened American and Western diplomacy on energy and the full range of issues that affect these countries and Western interests there. In Tashkent two months ago, I heard plenty of comments about the deterioration of US-Uzbek relations after the terrible events in Andijan in 2005, but no one talked about wanting the United States less engaged. On the contrary, those I spoke with wanted us to be more present and more involved, including domestic reform. In Azerbaijan last June, I heard complaints, too, but Azeris wanted more United States, not less. The same can be said of Kazakhstan, Kyrgyzstan, Georgia, Armenia and most, if not all of the countries in the region.

It was helpful that Secretary of State Hillary Clinton and Secretary of Defense Robert Gates visited the Caucasus last summer. Assistant Secretary of State for South and Central Asia Robert Blake and others have started to redevelop US dialogue with Central Asia. I would urge members of Congress and other senior officials in the Executive Branch to get us back in the game in this important region and to do so in a personal and sustained way. Go and hear out the region's complaints and concerns. Try to understand where the leaders and peoples of these countries are really coming from. Resist the temptation to judge in black versus white, good versus bad terms. Work to identify common interests and ways to advance them together. Building cooperation, drawing these countries more toward the West and toward more prosperous, peaceful and free futures will be better for all of us.

Mr. Chairman, thank you for holding this hearing. I look forward to what I am sure will be an interesting and useful discussion.

In: European Energy Security Options ... ISBN: 978-1-62081-814-5
Editors: R. E. Phillips and J. B. Cook

Chapter 7

RUSSIAN ENERGY POLICY TOWARD NEIGHBORING COUNTRIES*

Steven Woehrel

SUMMARY

The Russian oil and natural gas industries are key players in the global energy market, particularly in Europe and Eurasia. Another trend has been the concentration of these industries in the hands of the Russian government. This latter phenomenon has been accompanied by an authoritarian political system, in which former intelligence officers play key roles.

Russian firms have tried to purchase a controlling stake in pipelines, ports, storage facilities, and other key energy assets of European countries. They need these assets to transport energy supplies to lucrative western European markets, as well as to secure greater control over the domestic markets of the countries of the region. In several cases where assets were sold to non-Russian firms, Russian firms cut off energy supplies to the facilities.

Russia has also tried to build new pipelines to circumvent infrastructure that it does not control. Another objective Russia has pursued has been to eliminate the energy subsidies former Soviet republics have received since the fall of the Soviet Union, including by

* This is an edited, reformatted and augmented version of a Congressional Research Service publication, CRS Report for Congress RL34261, prepared for Members and Committees of Congress, from www.crs.gov, dated March 22, 2010.

raising the price these countries pay for natural gas to world market prices.

It is not completely clear whether the pursuit of Russian foreign policy objectives is the primary explanation for the actions of its energy firms. Few would disagree in principle that the elimination of subsidies to post-Soviet countries is a sound business decision, even if questions have been raised about the timing of such moves. Even the pursuit of multiple pipelines can be portrayed as a business decision. On the other hand, many countries of the region are concerned that Russia may use their energy dependency to interfere in their domestic affairs or to force them to make foreign policy concessions. Countries of the region also fear that by controlling energy infrastructure in their countries, Russian energy firms are able to manipulate the internal political situation by favoring certain local businessmen and politicians. However, the current global economic crisis has hurt Russia's energy firms and Russia's international clout, as energy prices have tumbled.

Bush Administration officials repeatedly criticized what they viewed as Russian efforts to use its energy supplies as a political weapon. The Obama Administration, like its predecessor, has urged European countries to reduce their dependence on Russian energy, but has said the United States is trying to cooperate with Moscow on the issue. The United States has strongly advocated the building of multiple pipelines from Central Asia and Azerbaijan to Europe. Members of Congress have expressed concern about the impact on European countries of their dependence on Russian energy. In the 111th Congress, committees have held hearings that have touched on the issue. Congress has also passed resolutions that refer to worrisome aspects of Russian energy policy.

INTRODUCTION

In recent years, Members of Congress, Bush and Obama Administration officials, and analysts have noted the dependence of many European countries on Russian energy. Some experts have expressed concern that Russia is using this energy dependence as part of a larger effort to limit the sovereignty and pro-Western orientation of vulnerable neighboring countries such as Ukraine, Moldova, and Georgia. In addition to bolstering the sovereignty of these countries, the United States has also had a vital interest in keeping strong ties with NATO and EU member states. Some observers believe that these relations could be harmed in the long term if many of these states became too dependent on Russian energy.

In 2008, the European Union depended on Russia for 23% of its natural gas consumption.[1] However, this figure conceals the fact that this dependence is unequally distributed. Some EU countries, many of them in central and eastern Europe, are dependent on Russia for most or all of the oil and natural gas they consume. For example, the Baltic states are entirely dependent on Russia for natural gas. Non-EU countries bordering Russia are also overwhelmingly or entirely dependent on Russian oil and natural gas.[2] Moreover, EU countries may become more dependent on natural gas supplied by Russia in the future, as deposits in the North Sea decline. In 2008, 40% of the EU's natural gas imports came from Russia and 33% of its oil imports.[3]

This report begins with a brief discussion of the Russian oil and gas industries, including their efforts to purchase energy infrastructure in central and eastern Europe and reduce energy subsidies to neighboring countries. A second section deals with the impact of Russian energy policy on neighboring countries, all of them formerly part of the Soviet Union, *de facto* or *de jure*, and all heavily dependent on Russian energy imports. Many of these countries are concerned about what they see as Russian efforts to manipulate that dependency to achieve political goals. A final section deals with US efforts to promote the energy security of these countries and on Congress's response to the issue.[4]

RUSSIA'S OIL AND GAS INDUSTRIES AND RUSSIAN FOREIGN POLICY

Russian oil and natural gas industries are very important players in the global energy market, particularly in Europe and Eurasia. Russia has by far the largest natural gas reserves in the world, possessing over 30% of the world's total. It is eighth in the world in oil reserves, with at least 10% of the global total. Another key trend has been the concentration of these industries in the hands of the Russian government in the past decade. This latter phenomenon has been accompanied by an authoritarian political system under the tight control of President Vladimir Putin, a former officer of the Soviet KGB intelligence service. Both the leadership of state-controlled Russian oil and gas firms and the Russian government are dominated by former members of the Russian intelligence service, now called the Federal Security Service (FSB), or are personally close to Putin, or both. In late 2007, Putin designated First Deputy Prime Minister Dmitri Medvedev (whose functions included overseeing Gazprom) as his successor. Medvedev was elected president of Russia in March 2008, in a vote viewed by many observers as not free and fair. Medvedev, although a close Putin associate, is not a former FSB officer. After Medvedev took office, Putin took the post of Prime Minister.

The personal and political fortunes of Russia's leaders are tied to the energy firms, as Russia's economic revival in the past decade has been due in large part to the massive revenues generated by energy exports. However, many experts believe that the Russian leadership's state-oriented approach may be counterproductive for Russia in the long run, as output growth in Russian oil and gas fields is stagnating. They say Russia's oil and gas industries will likely need foreign investment and expertise, more efficient management, as well as less government regulation and taxation. The global economic crisis hurt Russia's energy firms, Russia's economy, and Russia's international clout, as energy prices tumbled.

Gazprom

Since the collapse of the Soviet Union in 1991, the largest firm in Russia has been the state-controlled natural gas monopoly Gazprom. (The Russian government holds just over 50% of its shares.) It has a monopoly on gas pipelines in Russia. It controls about 85% of Russian gas production and over a quarter of the world's reserves of natural gas. Its impact within Russia is even more significant. It is the single largest contributor to the Russian government's budget, providing about 25% of tax receipts. It also controls banks, industrial holdings, farms, and media outlets.

Gazprom has been useful domestically to Russian leaders. It provides most of its production at a loss (at prices much lower than those in Europe) to Russian companies and consumers, who often cannot or will not pay, thereby helping to ease social pressures. In exchange for subsidizing Russian domestic consumers, Gazprom receives a virtual monopoly on exports to richer customers abroad. Two-thirds of Gazprom's revenue comes from European customers. Many experts say Gazprom needs to substantially increase domestic prices for gas if it is ever to become a viable business. Russia has agreed to gradually increase some domestic gas prices to bring them closer to world market levels by 2011.

The move was taken in response to EU criticisms of the price subsidy in negotiations over Russia's entry into the World Trade Organization (WTO).[5]

Gazprom's key current gas fields are in decline, its infrastructure is aging, and substantial investment will be needed just to maintain current, nearly stagnant production levels. Most foreign observers believe Gazprom could use foreign investment to provide expertise and capital. In late 2005 Russia abolished the "ring fence" that barred foreigners from owning Gazprom shares. Foreign ownership of Gazprom shares is modest; the largest foreign investor is the German firm E.ON, with a 6% stake.[6]

However, Gazprom's actions in other areas appear to point toward more state control and not toward more foreign investment. For example, in 2006, Gazprom bought out half of the interest of Shell and other foreign companies in the Sakhalin-2 gas field project at a very low price after the Russian government found alleged environmental problems in Shell's management of the project. Russia is also pressuring the Russian-British TNK-BP oil firm to hand over its Kovytka gas field in Siberia cheaply to Gazprom, after Gazprom refused to approve an export route for the gas.

Russian Oil Companies

In the 1990s, the Russian government did not have a large stake in Russian oil production. The major oil companies were controlled by politically well-connected businessmen, dubbed oligarchs. This policy changed in 2003, when Mikhail Khodorkovsky, head of the giant Yukos oil firm, was arrested on tax evasion and other charges. However, most observers believe that the real reason for the arrest was that President Putin believed Khodorkovsky was showing unacceptable signs of political independence from the Kremlin. Yukos was hit with government claims for back taxes and declared bankrupt. In 2004, its key oil field assets were bought very cheaply by the Russian state-owned oil firm Rosneft. The Yukos affair is not the only example of this trend. In 2005, Gazprom bought another major oil company, Sibneft, from oligarch Roman Abramovich, at a cut-rate price. The firm is now called Gazprom Neft. After the move, the Russian government controlled over 30% of Russia's oil production.[7]

In 2006, Putin said that the state would not take control of additional Russian oil companies. However, the strengthening of state control over the industry may be conducted by other means. The Russian government has placed pressure on foreign oil companies to sell their stakes in lucrative Russian oil fields to Russian state firms. The non-state oil firms that remain, such as Lukoil, are careful to retain close connections to Russian political leaders, in order to retain control of their businesses. Despite increasing direct and indirect state control, Russian oil firms continue to seek ties to foreign oil companies, provided that they are satisfied with a minority ownership stake.

In addition to increasing control over oil production, the government controls Russia's oil and refined product pipelines, through the state firm Transneft. This monopoly gives the Russian government leverage against Russian private firms, foreign investors and foreign countries, if needed. For example, Transneft is the largest shareholder in the Caspian Pipeline Consortium (CPC). The CPC pipeline carries oil from Kazakhstan to the Russian port of Novorossiysk on the Black Sea. This gives Russia a near monopoly on the transport of Kazakh oil destined for western markets. Other members include ExxonMobil and Chevron, which are involved in the exploitation of the Kazakh oil fields. In addition, Moscow has presented the Western oil companies with repeated financial demands, and threatened them with legal proceedings for allegedly unpaid taxes to Russia if they do not comply.[8]

Russia's Objectives: Exploiting Dependency or Just Good Business?

Through its energy firms, Moscow has pursued several objectives. Russia has tried to purchase a controlling stake in pipelines, ports, storage facilities, and other key energy assets of European countries. Russia needs these assets to transport energy supplies to Western European markets, as well as to secure greater control over the domestic markets of the countries of the region. In several cases where energy infrastructure was sold to non-Russian firms, Russia cut off energy supplies to the facilities.

Russian firms have attempted to buy energy infrastructure in western European countries, provoking unease in the EU. The EU has pressed Russia to open up its pipelines to western firms and to provide stronger protections for foreign investment in Russia's energy sector. Russia flatly rejected EU demands that it ratify the 1994 Energy Charter Treaty, which enshrines these principles. EU efforts to include the key components of the Energy Charter into a new Russia-EU Partnership and Cooperation Agreement (PCA) have also been stymied. The EU Commission has proposed an EU energy policy called "unbundling" that would prohibit energy-producing companies from owning distribution networks. It would also bar foreign companies from investing in EU distribution networks, unless that country permitted such investment in its own networks. Russia strongly criticized the Commission proposal. Indeed, Russian officials have said that participation of foreign companies in energy projects in Russia may be conditioned on an ownership stake for Russian firms of "downstream" assets in those countries.

EU countries have also been concerned about Russian efforts to coordinate export policy with other natural gas-producing countries such as Algeria, evoking the specter of a cartel, or "gas OPEC." Russia has also hinted that the bulk of Russian energy exports could be provided to Asia or even the United States instead of Europe in the future, once new pipelines to Asia are completed and new liquefied natural gas facilities are built.

It is unclear whether the EU can adopt an effective common policy on the Russian energy question. Central and eastern European countries within the EU want the EU to take a stronger stance against dependence on Russia for energy. On the other hand, energy companies and other influential voices in countries such as Germany, France, and Italy have forged close links with Moscow in order to secure access to Russian energy supplies and opportunities to invest in Russian energy projects. They assert that a mutual dependence exists between Russia and consuming countries. They note that roughly two-thirds of Russia's export revenues and half of its state budget comes from oil and gas exports. They insist that the current system of gas pipelines gives Russia little choice but to sell its gas exports to Europe.[9] Many European countries are less dependent on oil supplies from Russia, as they can draw on a more flexible global oil market, with the oil delivered by tankers.

Facing difficulties in securing control of energy infrastructure in central and eastern Europe, Russia has tried to bypass countries in the region entirely where possible. It is expanding the use of the Baltic Pipeline system and its oil terminal at the port of Primorsk and reducing the use of oil terminals in the Baltic states, such as Butinge in Lithuania and Ventspils in Latvia. In addition, Russia is considering oil pipeline projects involving Murmansk and other ports in northern Russia.

Russia's plans for natural gas pipelines have attracted the most attention. Gazprom has started work on the North European Gas Pipeline (NEGP), also known as Nord Stream, which would start transporting natural gas from Russia to Germany via a pipeline under the Baltic Sea as early as the end of next year, bypassing the states of central and eastern Europe. Ultimately, it would have a capacity of 55 billion cubic meters per year. In November 2007, Gazprom and the Italian firm ENI signed an agreement to build a "South Stream" gas pipeline that would run from Russia under the Black Sea through the Balkans, with branches to Austria and Italy. Bulgaria, Serbia, Croatia, and Hungary have also signed on to the project. Gazprom hopes to complete South Stream in 2015. South Stream would also bypass Belarus, Ukraine, Poland, and other central European countries. In May 2009, ENI and Gazprom agreed to expand South Stream's capacity to 63 bcm per year.

Another possible natural gas pipeline project is Yamal-Europe 2. This long-proposed pipeline, which would parallel a currently operating one that has a capacity of 33 bcm, would run through Belarus and Poland, bypassing Ukraine. The pipeline is unlikely to be built, as the Russian government and Gazprom have rejected Belarusian proposals to reactivate the Yamal-Europe 2 plans.

By seeking a range of transit routes through the region, Russia may be trying to reduce the leverage that transit countries, including those in central and eastern Europe, have in negotiations with Russian energy firms. Experts note that the capacity of these new routes, if built, would likely outstrip Russia's capacity to produce oil and gas to fill them, allowing Russia to allocate scarce production to "favored" transit countries. Russia may also be trying to reduce the attractiveness of other routes for oil and gas pipelines from Azerbaijan and Central Asia to Europe and Asia that would bypass Russia.[10]

Another objective Russia has pursued has been to eliminate the energy subsidies former Soviet republics have received since the fall of the Soviet Union, including by raising the price these countries pay for natural gas to world market prices.[11] These actions may be seen as paralleling the reduction of subsidies to Russian domestic consumers. However, Russia has also used the withdrawal of price subsidies and the unpaid energy debts of countries in the region as leverage to secure key energy infrastructure in those countries.

It is not clear whether the pursuit of Russian foreign policy objectives is always the primary explanation for the actions of its energy firms. Few would disagree in principle that the elimination of subsidies to post-Soviet countries is a sound business decision, even if questions have been raised about the timing of such moves. In support of their actions, Russian leaders point to the fact that Russian allies such as Armenia and Belarus have also been subject to energy price hikes. The pursuit of multiple pipelines can also be portrayed as a business decision, although some analysts disagree about its wisdom. They assert that Russia would do better to invest more in boosting production rather than building pipelines that it may not be able to fill as its current oil and gas fields decline.

Many countries of central and eastern Europe are concerned that Russia may use their energy dependency to interfere in their domestic affairs or force them to make foreign policy concessions. Gazprom's increases in energy prices to Georgia and Ukraine came after elections in 2004 brought to power pro-Western leaders in what were termed respectively the “Rose” and “Orange”

Revolutions, in reference to their campaign symbols. Analysts asserted that Russian leaders feared so-called “color revolutions” elsewhere in the former Soviet countries that could reduce Russia’s influence, and even perhaps threaten Russia’s own authoritarian regime. Countries of the region also fear that by controlling energy infrastructure in their countries, Russia is able to manipulate the internal political situation by favoring certain local businessmen with participation in local business ventures of Gazprom or other Russian energy firms. These businessmen are in a position to assume a powerful political role themselves or bribe politicians to do Moscow’s bidding.[12]

Critics of Russian policy say Moscow’s motives become even clearer when viewed in the context of other actions to apply pressure to neighboring states. For example, Moldova’s economy was seriously harmed by a wine import ban Russia imposed, ostensibly for health reasons, from 2005 to 2007. Georgia has been hurt by Russian immigration restrictions and a wine ban. Both countries have also faced problems with Russian support for breakaway regions on their territories.

Russia’s August 2008 military assault on Georgia and its subsequent recognition of the independence of breakaway regions South Ossetia and Abkhazia sent a powerful signal to Russia’s neighbors about the costs of angering Moscow.

NATO and EU member Estonia suffered from cyberattacks that may have been instigated by Russia, in the wake of a controversy over the removal of a Soviet-era military statue from Estonia’s capital in April 2007. Russia’s critics also point to Russia’s “National Security Strategy to 2020,” released in May 2009, which states that “the resource potential of Russia” is one of the factors that has “expanded the possibilities of the Russian Federation to strengthen its influence on the world arena.”[13]

While Moscow is often charged with using energy policy to pursue foreign policy goals, it may also at times use foreign policy issues to benefit its energy firms. In January 2008, the Russian natural gas monopoly Gazprom reached an agreement with Serbia to buy NIS, the Serbian national oil company, at what some observers believed to be a below-market price. Gazprom may have been able to achieve this in part due to then-Serbian Prime Minister Vojislav Kostunica’s appreciation for Moscow’s opposition to independence for Serbia’s Kosovo province.

SELECTED RECENT CASES

The countries discussed in this report have all faced the impact of Russian energy policy. All are heavily or entirely dependent upon Russia for their natural gas and oil imports. They face common issues of cost, reliability of supplies, and Russian efforts to control downstream infrastructure such as pipelines, refineries, and domestic distribution networks. The countries differ in their geopolitical orientation. Ukraine and Moldova seek eventual EU membership, while Georgia seeks to join the EU and NATO. The Baltic states are EU and NATO members, but Russia maintains a strong hold on their energy sectors. On the other hand, Belarus is Russia's most loyal supporter in Europe, while Armenia is a strong ally of Moscow in the south Caucasus region.

Ukraine

Although it possesses modest oil and natural gas reserves of its own, Ukraine is dependent upon Russia for most of its oil and natural gas, both from Russia's own oil and natural gas fields and from Russian-controlled pipelines from Ukraine's suppliers in Central Asia, especially gas from Turkmenistan. In 2006, half of Ukraine's energy consumption came from natural gas. Sixty-six percent of Ukraine's domestic natural gas consumption came from Russia. In 2006, Russian oil imports accounted for 78% of Ukraine's oil consumption. [14] Most Ukrainian homes are heated by natural gas. Ukraine's steel and other heavy industries, which play a key role in Ukraine's exports, are highly inefficient users of energy. However, Ukraine's vulnerability to Russian pressure has been mitigated by the fact that the main oil and natural gas pipelines to central and western Europe transit its territory. Ukraine's gas pipeline system has a capacity of 120 billion cubic meters per year. Ukraine owns the sections of the pipelines that run through its territory as well as large gas storage facilities. Ukraine has received transit fees from Gazprom, paid partly in gas and partly in cash. According to a 2006 report of the International Energy Agency, 84% of Russia's natural gas exports and 14% of Russian oil exports pass through Ukraine.[15]

Energy issues have played a key role in Russian-Ukrainian relations since the breakup of the Soviet Union in 1991. Russian firms supplied energy to Ukraine at prices far below market rates. In the early 1990s, these firms cut off supplies to Ukraine at times due to unpaid energy debts. Energy sales have been conducted by non-transparent intermediary institutions, offering the elites of both countries opportunities to profit.

Since achieving independence in 1991, Ukraine has tried to strike a balance between improving ties with the West, including nominal support for Euro-Atlantic integration, while not offending Moscow. However, in 2005 Viktor Yushchenko was elected President of Ukraine, overcoming the previous regime's attempts at electoral fraud, in what was termed the "Orange Revolution." Russian leaders, who had strongly backed his opponent, Prime Minister Viktor Yanukovych, reacted angrily to Yushchenko's victory. Yushchenko, Ukraine's first clearly pro-Western leader, said Ukraine would conduct serious reforms so that it could join NATO and the European Union as soon as it was ready.

Soon after Yushchenko took office, Gazprom started to demand a sharp increase in the price of natural gas that it supplied to Ukraine. When Ukraine rejected this proposal, Russia cut off natural gas supplies to Ukraine on December 31, 2006. Ukraine then diverted to its own use some of the gas that Gazprom intended for European customers. After western European governments protested sharply, Gazprom resumed gas deliveries on January 2. Two days later, the Russian government and Gazprom reached an agreement with Ukraine for ensuring gas supplies to Ukraine. The agreement called for gas to be purchased by Ukraine through an intermediary firm, RosUkrEnergo. This firm paid for gas from Central Asia at a price lower than market levels, and adds gas from Russia at market prices. The agreement also provided for higher transit fee payments to Ukraine (now entirely in cash rather than partly in gas).

Perhaps more troubling for Ukraine, the accord called for the creation of UkrGazEnergo, a joint venture between RosUkrEnergo and the Ukrainian state-controlled gas firm Naftogaz that grants the former access to one-half of Ukraine's domestic market. Ukraine's intelligence service reportedly believes the owners of RosUkrEnergo are using their control over energy supplies to secure ownership of energy intensive industries such as fertilizer plants and a titanium plant.[16]

Some analysts are concerned about possible involvement of organized crime groups in RosUkrEnergo, as well as corrupt links with Russian and Ukrainian officials. The US Justice Department reportedly investigated the firm.[17] Nominally, Gazprom owns 50% of RosUkrEnergo, Ukrainian businessman Dimitry Firtash owns 45%, and another Ukrainian businessman owns 5%. In 2005, Ukranian Prime Minister Yuliya Tymoshenko, who was an important player in the natural gas industry in the 1990s, called for the elimination of

RosUkrEnergo as a middleman. She was dismissed by Yushchenko in September 2005, in a move that some observers believed was aimed in part at appeasing Gazprom and its supporters within the Ukrainian government.

Yanukovych's party won Ukraine's March 2006 parliamentary elections, and Yanukovych once again become Prime Minister. Gazprom's discussions with the Yanukovych government in late 2006 went more smoothly than those of the previous year. In 2007, Russia and Ukraine agreed on a moderate increase in the natural gas price. The two sides agreed to gradually increase the price of Russian natural gas to Ukraine over the next five years, until it reaches the world market price. Some observers have seen Gazprom's tough attitude toward Ukraine in the 2005 negotiations and its relatively benign stance in 2006 as evidence that Russia has manipulated the gas issue to undermine Yushchenko. In September 2007, Putin appeared to verify this view when he said that Russia had no desire to provide cheap energy to "Orange" forces.[18]

On September 30, 2007, Ukraine held closely contested parliamentary elections. On October 2, as the vote count showed a narrow victory by "Orange" parties, Gazprom announced that it would reduce gas supplies to Ukraine, if Ukraine did not pay outstanding debts to Gazprom by the end of the month. Gazprom officials hinted that Ukraine's energy debts could be solved if it turned over shares in the gas pipeline system to Gazprom as payment. However, the crisis was resolved when the Ukrainian government agreed to provide gas in Ukrainian storage facilities as payment. Gazprom and the outgoing Yanukovych government agreed to a sharp increase in the natural gas price, but still well short of world market levels. Nevertheless, the percentage increase was double that given to neighboring Moscow ally Belarus.

In December 2007, Yuliya Tymoshenko was elected by the new Ukrainian parliament as Prime Minister. She vowed to remove RosUkrEnergo and UkrGazEnergo from Ukraine's gas market. Gazprom reduced gas supplies to Ukraine by 50% on March 3-5, 2008, over disagreement on the price Ukraine should pay for gas delivered in January and February 2008. Ukrainian gas company officials warned that they might divert gas intended for Western Europe to offset Gazprom's supply cut. The two sides reached agreement on March 5 and supplies were restored. On March 12, the two sides agreed to eliminate UkrGazEnergo from the domestic gas trade, but gave Gazprom direct access to the most lucrative part of Ukraine's domestic market—supplies to large enterprises. The agreement said the fate of RosUkrEnergo would be determined by future negotiations. However, Tymoshenko appeared dissatisfied with parts of the accord. She remained determined to eliminate

RosUkrEnergo as soon as possible. The Ukrainian government made significant unilateral changes to the agreement, including barring RosUkrEnergo from the profitable practice of re-exporting Central Asian gas from Ukraine to other markets.

On January 1, 2009, the state-controlled Russian natural gas firm Gazprom halted gas supplies to Ukraine after the two sides failed to reach agreement on several issues, including a debt allegedly owed by Ukraine to Gazprom and the price that Ukraine would pay for gas supplies for 2009. The cutoff was supposed to affect only supplies for Ukraine; Russia continued to send gas through Ukraine destined for other European customers. However, within a few days, Russia accused Ukraine of diverting these supplies for its own use, and by January 6 cut off all deliveries through Ukraine to the rest of Europe. The European Union sharply criticized the cutoff, calling for a rapid resumption of supplies, but refused to take sides in what it termed a commercial dispute.

Many large European countries (including Ukraine) did not suffer greatly from the cutoff, despite frigid temperatures, due in part to substantial amounts of gas in underground storage facilities. However, some countries, particularly in the Balkans, were hit hard. Negotiations between Russia and Ukraine repeatedly broke down, with each side accusing the other of bad faith and trying to enlist European support for its views. Finally, on January 18, Russia and Ukraine reached an agreement, and gas supplies to Europe resumed on January 20th.

According to the agreement, RosUkrEnergo was eliminated as a middleman in the gas trade. In the first quarter of 2009, Ukraine was charged more than double what it paid in 2008 for gas. However, Ukraine was able to avoid this price increase by using supplies it had in storage until the gas price declined in the second quarter, as the sharp drop in oil prices was reflected in natural gas prices. The poor financial situation of Naftogaz and its customers, as well as the devastating impact of the global economic crisis on Ukraine, threatened to lead to further gas crises, as Gazprom threatened to cut off supplies again if it was not paid the full amount it was owed each month. However, Ukraine was able to find the money to make the payments and a crisis was avoided in 2009. In November 2009, Russia and Ukraine reached a revised gas deal that confirmed Moscow's earlier pledges that Russia would not impose fines on Ukraine for not taking enough gas from Russia.

Russia has pursued a long-standing goal of ownership of Ukraine's natural gas pipelines and storage facilities, as well as its local gas distribution network. In February 2007, Putin announced that he and Prime Minister Yanukovych had agreed on joint Russian-Ukranian control of Ukraine's natural gas assets, in exchange for a Ukrainian stake in Russian natural gas fields. However, this statement provoked a strongly negative reaction in Ukraine, and the parliament quickly approved a law banning any transfer of control of the pipelines by a vote of 430-0. Russia has tied possible support for upgrading Ukraine's pipeline network to greater Gazprom ownership of Ukraine's pipeline system.

In July 2009, the EU Commission brokered an agreement under which the European Bank for Reconstruction and Development (EBRD) and World Bank agreed to provide loans to help Ukraine modernize its gas pipeline system. Ukraine would be required to create greater transparency in how its gas sector is run. In the short term, $300 million in EBRD loans would be aimed at helping Ukraine improve its natural gas storage facilities and other infrastructure. In 2010, the EBRD may loan Ukraine another $450 million for Ukraine to reform its gas sector. The World Bank is willing to loan Ukraine up to $500 million for structural reforms, including of the gas sector.

In March 2009, Prime Minister Putin sharply criticized the initial announcement of the EUUkraine gas sector agreement as "ill-considered and unprofessional." Russian officials said that any agreement about Ukraine's pipelines should include Russia. Russia's anger may be caused by fears that the EU plan could foil Gazprom's own efforts to secure control of Ukraine's pipelines and Ukraine's domestic gas distribution network. The EU plan soon stalled, due to Ukraine's unwillingness to undertake reforms during the impending Ukrainian presidential election campaign.

Balked in its efforts to control Ukraine's pipelines for the time being, Russia is working on ways to bypass Ukraine's gas transport system, at least in part. It is developing new energy export routes through the Baltic Sea (the Nord Stream pipeline) and the Balkans (the South Stream pipeline) to western Europe beginning as early as the end of 2011 in the case of Nord Stream and 2015 in the case of South Stream. If successful, these efforts could reduce Ukraine's leverage over Russia on energy issues. Such a development could have broader consequences. Russia could feel it would have a freer hand to put greater pressure on Ukraine on other issues. Key western European countries could feel they have less of a stake in Ukraine's future, if they, like Russia, were no longer dependent on Ukrainian gas transport infrastructure.

The victory of Viktor Yanukovych in Ukraine's 2010 presidential elections may result in a change in Ukraine's energy relationship with Russia. Yanukovych is expected to pursue a more pro-Russian foreign policy than Yushchenko did. President Yanukovych has criticized the current gas agreement between Russia and Ukraine, saying that the gas is too expensive.

He wants Russia to return to its old policy of providing gas to Ukraine at subsidized prices, in exchange for Russia's gaining a large minority stake in Ukraine's pipeline system. Companies from western European consumer countries would also gain shares in the system, according to Yanukovych. However, it is not completely clear whether Gazprom, in its present financial difficulties, would be willing to cut gas prices for Ukraine. Secondly, the Ukrainian parliament would have to pass a law permitting the deal, which could prove highly controversial. Many in the parliament could accuse Yanukovych of breaking the law and selling off the country's chief strategic asset to Moscow.

Yanukovych may also revive the role of non-transparent middlemen in the gas trade, given the presence of persons in his government who were closely associated with RosUkrEnergo. It is unclear whether Ukraine will move forward with the EU-Ukraine gas sector reform plan, which calls for greater transparency in the Ukrainian gas system.

Moldova

Moldova is the poorest country in Europe, according to the World Bank. It is entirely dependent upon Russia for its energy resources, and also as a market for the wine and agricultural products that are its main exports. In 2005, Russia restricted wine and other agricultural imports from Moldova, allegedly over health concerns, dealing a very heavy blow to the country's economy. Russia has stalled on implementing pledges to end its embargo on Moldovan wine, still citing health concerns. In addition, Russia has supported a breakaway regime in the Transnistria region of the country, including by deploying 1,500 troops there.

In part due to its vulnerable position, Moldova has tried to balance ties between Moscow and Western countries since securing independence from the Soviet Union in 1991. However, Moldova's Communist leaders increasingly sought greater engagement with the West, irritating Russia. Perhaps even more irksome to Moscow, in 2005, Moldova, with EU help, began to tighten its customs policies to stop profitable smuggling operations from Transnistrian territory. Powerful groups in Ukraine and Russia have profited from the Transnistria regime's activities. Russia provides subsidies to Transnistria, which include grants and loans as well as subsidized energy. In return, Russian firms have received stakes in Transnistrian businesses.[19] All but a small fraction of Moldova's current debt to Gazprom for gas is owed by Transnistria.

Russia has pressured Moldova on the issue of energy supplies. On January 1, 2006, the Russian government-controlled firm Gazprom cut off natural gas supplies to Moldova, after Moldova rejected Gazprom's demand for a doubling of the price Moldova pays for natural gas. Gazprom restored supplies on January 17, when Moldova accepted a slightly smaller increase. Moldova also agreed to give Gazprom, already the majority shareholder, Transnistria's 13% stake in MoldovaGaz, which controls Moldova's natural gas pipelines and other infrastructure (Moldova had earlier ceded majority control to Gazprom in exchange for settling Moldova's gas debts). As a result of the agreement, Gazprom holds 63.4% of MoldovaGaz's shares and has control of Moldova's domestic gas infrastructure. Gas supplies to Moldova were cut off during the January 2009 Russia-Ukraine gas crisis, and were restored afterwards. The global economic crisis has resulted in a drop in the price of Moldova's gas supplies. The establishment of a pro-Western government in Moldova after the victory of anti-Communist parties in 2009 parliamentary elections has not led so far to dramatic changes in the country's energy relations with Moscow.

Georgia

Georgia began to follow a clearly pro-Western orientation after the "Rose Revolution" of November 2003, which swept out of power political forces with close ties to Russia after they had tried to use electoral fraud to win legislative elections. Mikhael Saakashvili won presidential elections in early 2004. Georgia is seeking NATO membership. Georgian-Russian relations deteriorated in the wake of the Rose Revolution. Russia has many ways to pressure Georgia, including supporting the breakaway regions of South Ossetia and Abkhazia, and disrupting economic ties between Georgia and Russia, including in the energy sphere.

In late 2005, Gazprom announced substantial increases in the price of gas shipped to Georgia. In the winter of 2005-2006, unknown saboteurs bombed gas pipelines in Russia, temporarily cutting off supplies to Georgia. Gazprom announced in November 2006 that it would cut off gas supplies to Georgia by the end of the year unless Georgia agreed to a 100% price hike or sold its main gas pipeline to Gazprom.

However, Georgia's geographical position neighboring energy-rich Azerbaijan has allowed it to counter Russian pressure more effectively than other countries. Georgia is a transit state for a pipeline completed in mid-2006 carrying one million barrels per day of Azerbaijani oil to the Turkish port of Ceyhan (the Baku-Tbilisi-Ceyhan or BTC pipeline). Another pipeline completed in early 2007 initially carries 2.2 billion cubic meters of Azerbaijani natural gas to Georgia and Turkey, lessening their dependence on Russia as a supplier. Another pipeline carries oil from Baku to the Georgian port of Supsa.[20]

The August 2008 Russian military assault on Georgia resulted in temporary interruptions of some oil and gas pipeline shipments, but these were soon resumed. Gazprom also continues to supply Georgia with natural gas, in part because a pipeline that delivers gas to Armenia passes through the country. Nevertheless, the conflict demonstrates Russia's ability to disrupt pipeline routes through Georgia to European customers, by military means if necessary. Georgian officials believe that the gas Georgia would receive from participating in the US-sponsored Nabucco pipeline project, if it is built, would meet all its natural gas needs.

Baltic States

The Baltic states of Lithuania, Latvia, and Estonia are members of NATO and the EU. They have often had difficult relations with Moscow. About 90% of their oil comes from Russia, and 100% of their natural gas. They faced Soviet energy supply cutoffs in the early 1990s, as they were trying to achieve independence and shortly thereafter. They pay world market prices for their energy supplies. In the past few years, the main concern in the Baltic states has been Russian efforts to increase control over the energy infrastructure in their countries. Gazprom has a large equity stake in domestic natural gas companies of each of the three Baltic countries.[21] When Russian takeover efforts have failed, Russia has cut off energy supplies to Baltic energy facilities. One striking case involves the Mazeikiai oil complex in Lithuania. Mazeikiai includes a large refinery, the Butinge maritime terminal, and a pipeline. It is the largest enterprise in Lithuania (accounting for about 10% of Lithuania's GDP) and provides vitally needed tax revenue. In 1999, the US firm Williams International bought a large stake in Mazeikiai and also received the operating rights. In response, the Russian oil firm Lukoil, which supplied the oil to the refinery, slowed deliveries to a trickle, making Mazeikiai unprofitable. This led Williams, which had financial problems of its own, to sell its stake to Yukos in 2002.

Under Yukos, the refinery became profitable again. However, when Yukos later fell afoul of Russian authorities, and was driven into bankruptcy, Yukos attempted to sell its stake in Mazeikiai. The Polish oil firm PKN Orlen agreed to buy out Yukos, despite an effort by the Russian government-controlled oil company Rosneft to purchase the refinery. At the end of July 2006, the Russian government-owned oil transport company Transneft announced that the part of the Druzhba oil pipeline that supplies Mazeikiai was temporarily shutting down for repairs following an oil leak. Transneft later said that it would not reopen the pipeline, due to its unprofitability. Transneft blocked Lithuania's efforts to secure supplies from Kazakhstan through Transneft's pipelines. Critics charged that once again Russia was manipulating energy supplies to punish Lithuania for seeking to diversify ownership in its energy sector.[22] In February 2010, Lithuanian Prime Minister Andrius Kubilius said that a 25% stake in Mazeikiai could be sold to a Russian oil firm in exchange for oil supplies.

Another example of a Russian company using its control over energy supplies in an effort to strong-arm a Baltic country into handing over key infrastructure occurred in January 2003. The Russian government-controlled Transneft oil pipeline company cut off all oil shipments to the Latvian oil terminal at the port of Ventspils, after having decreased shipments in late 2002. The move was a large blow to Latvia, as Ventspils has been important to Latvia's economy. Transneft diverted the oil shipments to its own Baltic Pipeline System and the Russian port of Primorsk, which it controls. Transneft claims that there is no demand for using Ventspils, a claim viewed with skepticism by outside observers. Most saw the move as a power play by Transneft to secure a controlling share of the firm Ventspils Nafta, which operates the oil terminal.[23] Estonia has also felt the effects of Russian pressure on its energy supply. On May 2, 2007, Russia's state railway monopoly halted delivery of oil products and coal to Estonia in the midst of a political furor over the relocation of a Soviet war memorial statue from a square in central Tallinn, Estonia's capital.[24]

Belarus

Belarus, under the authoritarian leadership of President Aleksandr Lukashenko, has been Moscow's most loyal ally in Europe since the collapse of the Soviet Union. Belarus's unreformed, largely Soviet-style economy is heavily dependent on cheap Russian natural gas and oil. Gazprom long supplied Belarus with energy at Russian domestic prices, providing a large indirect subsidy to the Lukashenko regime.

However, in 2006, it appeared that Russia had decided to reduce its subsidies to Lukashenko. In late 2006, Gazprom strongly pressured Belarus to sell to it control of the Beltransgaz natural gas firm (which controls the pipelines and other infrastructure on Belarusian territory) and other key Belarusian energy firms, or face the quadrupling of the price Belarus would pay for Russian natural gas. Gazprom threatened a cutoff in supplies on January 1, 2007, if Belarus did not agree to pay the higher price. Just hours before the deadline, however, the two sides reached an agreement that averted a gas shutoff. Belarus agreed to pay more than double what it paid in 2006. Belarus's natural gas prices were scheduled to rise steadily, reaching world market levels in 2011. Belarus also agreed to sell Gazprom a majority stake in Beltransgaz.

After settling the gas dispute, the two countries were soon embroiled in conflict over oil supplies. In addition to receiving cheap natural gas, Belarus has also benefitted from inexpensive and duty-free crude oil supplies that are processed at Belarusian refineries. Belarus then sold the bulk of these refined products to EU countries at a hefty profit. In January 2007, Russia moved to sharply reduce this subsidy to the Belarusian economy. Russia imposed a tariff on oil exports to Belarus. Belarus retaliated by increasing transit fees for Russian oil supplies to Western Europe. When Russia refused to pay, Belarus cut off oil supplies to Western European countries, angering their governments. Belarus and Russia ended the crisis by agreeing that Belarus would raise its export duty on crude and refined oil products to Western Europe to match that imposed by Russia.

Russia would then exempt Belarus from most of the new Russian oil export duty. The drop in this subsidy from Moscow caused Lukashenko to cut some government spending and to look to Western banks for loans.

Another oil crisis occurred in January 2010. Russia demanded a revision in the oil export arrangements between the two countries. Russia wanted to supply a fixed amount of oil duty-free for Belarus's domestic use, but any amounts over that would be subjected to the full duty. In addition to a reduction of the Russian subsidy to Belarus, the plan would also put Russia's own refineries on a more equal footing with the subsidized Belarusian ones. Russian Deputy Prime Minister Igor Sechin (who is also chairman of the state-owned Rosneft oil company) said that if Belarus wants better terms it will have to sell equity stakes in its refineries to Russian firms. In order to put pressure on Belarus, Russia cut its supplies to Belarus by as much as 70% on January 23. Faced with a shutdown of its refineries, Belarus quickly agreed to a modified version of Russia's terms on January 27. It is estimated that the new oil supply deal will cut Russia's subsidy to Minsk by about $1.8 billion. Some analysts believe that it may make Belarus's refineries unprofitable. The crisis did not have an impact on oil supplies to Western Europe.[25]

Belarus continues to ask Russia to moderate future energy prices. Belarus paid about $148 per bcm for gas in the first quarter of 2009, far below that charged to Ukraine. Belarus also wants Moscow to increase the transport fees it pays to Belarus for use of its pipelines to Europe, so far without success. To hedge his bets, Lukashenko has called for diversification of Belarus's energy supplies by imports from such countries as Venezuela, Iran, and Azerbaijan, as well as development of coal reserves and a nuclear power plant within Belarus.

Belarus has also explored receiving oil supplies from Kazakhstan. The world economic crisis has eased Belarus's energy burden somewhat. Gas prices paid by Belarus since the second quarter of 2009 have dropped due to the fact that market rates for gas are tied to those for crude oil, which have plummeted in recent months.

Armenia

Armenia and Russia have close political and military ties, in large part due to Armenia's desire for support in its struggle with Azerbaijan over the Nagorno-Karabakh region of Azerbaijan. However, in early 2006, Russia informed Armenia that it would sharply increase the price it would have to pay for gas.

In May 2006, Armenia agreed to relinquish various energy assets to Russian firms as partial payment for this price increase. Some critics have alleged that Russia now has virtual control over Armenia's energy supplies.

In October 2006, Armenian officials announced that Gazprom would assume effective management control of an Iranian-Armenian gas pipeline. According to some experts, this acquisition may indicate Russia's intent to block use of Armenia as a pipeline route independent of Russian control. Some of the gas will be used to generate electricity for Iran and Georgia, but the remainder may satisfy all Armenia's other consumption needs, removing its dependence on Russian gas transported via Georgia.[26]

US Policy

The United States has criticized what it has viewed as Russian efforts to use its energy supplies as a political weapon. During the January 2006 natural gas standoff between Russia and Ukraine, State Department spokesman Sean McCormack criticized Russia for using "energy for political purposes." He stressed that while the Administration supported a gradual increase in prices to market levels, it disagreed with a "precipitous" increase and cutoff. Secretary of State Condoleezza Rice likewise on January 5 stated that Russia had made "politically motivated efforts to constrain energy supply to Ukraine."[27] In May 2006, Vice President Dick Cheney criticized Russia's tactics of "supply manipulation or attempts to monopolize transportation" against vulnerable countries in the region as "blackmail" and intimidation."[28]

In testimony before the Commission on Security and Cooperation in Europe in June 2007, Bush Administration officials cast doubt on the reliability of Russian oil and natural gas supplies to Europe and criticized Moscow's "nationalistic interventions in its energy sector."[29] On October 23, 2007, Secretary Rice noted that "we respect Russia's interests, but no interest is served if Russia uses its great wealth, its oil and gas wealth, as a political weapon, or that if it treats its independent neighbors as part of some old sphere of influence."[30]

A State Department spokesman reiterated US opposition to the use of energy supplies as a political weapon after a March 3-5, 2008, gas supply incident between Gazprom and Ukraine.

The Obama Administration has also expressed concerns about European energy security. On January 22, 2009, after the resolution of the second major Russia-Ukraine gas crisis, a State Department spokesman said the conflict "underscores the need for transparent, market-oriented arrangements for the sale and shipment of natural gas and the importance of diversifying energy supplies.[31]

In April 2009, the Obama Administration appointed Richard Morningstar as State Department Special Envoy for Eurasian Energy.

While the United States has been active on the issue of Eurasian energy supplies, skeptics might argue that key US interests are not at stake, given the fact that the problem is one of European dependence, not that of the United States.

Moreover, the EU and other European countries will have to be the main players in finding a solution (if they have the will to do so), with the United States playing a secondary role.

US "Pipeline Diplomacy"

The United States has urged European countries to diversify their energy, supplies, reducing their dependence on Russia. The United States has strongly advocated the building of multiple pipelines to supply energy from Central Asia and Azerbaijan to Europe. These projects include the Baku-Tbilisi-Ceyhan (BTC) pipeline), which carries 1 million barrels per day of Azerbaijani oil to the Turkish port of Ceyhan. Another important project supported by the United States has been the South Caucasus Gas Pipeline (SCGP), which taps Azerbaijan's Shah Deniz gas field.

In the longer term, through about 2020, the United States supports expanding the SCGP. The United States also strongly supports the Nabucco pipeline, an EU-sponsored project that would supply natural gas from Central Asia and Azerbaijan to Europe through Georgia, Turkey, Bulgaria, Romania, Hungary and Austria. Nabucco received a further boost on July 13, 2009, when Austria, Hungary, Romania, Bulgaria, and Turkey signed an intergovernmental agreement on the project. It is hoped that work on the pipeline could begin in 2011, with the first gas supplies available by 2015. The full capacity of the pipeline, planned at 31 bcm, would be reached a few years later.

Another US-backed proposal is the Italy-Turkey-Greece Interconnector (ITGI) gas pipeline. The connection of the Turkish and Greek gas pipelines was completed in November 2007. The Greek and Italian gas transport systems are scheduled to be connected by 2012. The pipeline will have a capacity of 8 bcm per year, substantially smaller than Nabucco or South Stream. ITGI may be linked to Nabucco, if it is built.

Turkey plays a central role in all of these pipeline projects. Turkey limits shipping in the Bosporus Strait due to environmental concerns. This reduces the potential for tankers to use the Black Sea to ship oil and gas to European markets. Therefore, supplies from the Caucasus and Central Asia must cross Turkey's territory or transit across the Black Sea from the Caucasus in the east to the Balkans in the west. The United States has supported the American Macedonian Bulgarian Oil pipeline (AMBO), another Bosporus bypass project. It could supply Caspian oil from the Bulgarian Black Sea port of Burgas through Macedonia to Albania's Adriatic port of Vlore. It is expected to be completed in 2011.

Turkey appears to be using its prime geographic location as bargaining leverage with both Russia and the EU, in hopes of making Turkey a regional energy hub. After signing the intergovernmental agreement on Nabucco in

July 2009, Turkey agreed in August to allow Russia to use Turkey's territorial waters for South Stream. In exchange, Russia will provide support for a trans-Anatolian oil pipeline crossing Turkey from the Black Sea to the Mediterranean. Russia and Turkey will also discuss expanding the Blue Stream natural gas pipeline between Russia and Turkey. In addition, Russia also agreed to build a nuclear power plant in Turkey.

Another proposed gas pipeline is White Stream. It would provide gas from the Caspian through a pipeline under the Black Sea from Georgia to Crimea, in Ukraine. The pipeline would have to pass over Russia's Blue Stream pipeline to Turkey on the seabed.[32] In April 2009, Georgia and the White Stream Pipeline Company signed a memorandum of understanding on the project. However, it is unclear whether the pipeline can find financing in the current economic climate, or a source of supply.

Bush Administration officials criticized Nord Stream, which would traverse the Baltic Sea floor, supplying Germany and other western European countries with natural gas, and bypassing the central and eastern European countries through which the main current pipelines run. The US supported discussions by Poland, the Nordic countries, and the Baltic states on alternatives, including delivering Norway's expanding gas production to northern Europe via Danish pipelines and by developing liquefied natural gas terminals in Poland and the Baltic states.[33] Bush Administration officials also criticized South Stream, saying that it will not reduce dependence on Russian supplies.

However, Germany strongly supports Nord Stream, as it is looking for a dependable source of natural gas, particularly after the natural gas crises between Russia and Ukraine, which briefly interrupted supplies to central and western Europe in 2006 and 2009. Gazprom owns 51% of Nord Stream, while two German firms and a Dutch firm, own the rest. A French company has also agreed to join the consortium. Former German Chancellor Gerhard Schroeder is chairman of the Nord Stream consortium. On the other hand, the Baltic countries, Poland and Ukraine have expressed concerns about Nord Stream, fearing that it will give Moscow more leverage on energy issues with them. After delays due to environmental concerns by Sweden, Finland, and Denmark, through whose waters the pipeline would pass, Nord Stream is expected to begin laying the underwater segments of the pipeline later this year, with hopes that the pipeline will begin operation in late 2011.

The Obama Administration has been less critical of Nord Stream and South Stream than the Bush Administration. Part of the change in tone may be due to an effort to "reset" ties with Russia that were frayed during the Bush years. US Eurasian energy envoy Richard Morningstar has denied that the United States and Russia are involved in a "great game"—that is, a geopolitical struggle—for Central Asian energy supplies. He has said that the United States should at least try to work with Russia on the issue. In contrast to statements by Bush Administration officials, Morningstar has said that the United States does not oppose Nord Stream and South Stream; that the United States does not see Nabucco as being in competition with South Stream; and that it is possible that Russia could provide gas for Nabucco.[34]

The United States has advocated extending an existing oil pipeline that currently runs from the oil terminal at Odesa in Ukraine to Brody, on the Polish border. This pipeline could then be extended to Gdansk in northern Poland. At present, however, the Odesa-Brody pipeline runs in the reverse direction, pumping Russian oil to Odesa.

Obstacles to US-Supported Pipelines

These US efforts at "pipeline diplomacy" face challenges. The success or failure of these projects will likely depend more on whether private energy firms find them profitable than on US diplomatic skill and energy. A particular concern is whether there is enough oil and natural gas to supply the various pipelines. Russian-supported pipelines appear to have the upper hand because they have supplies available. Their potential profitability makes them tempting even to central and eastern European countries that are trying to reduce dependence on Russia. In addition, Moscow can offer them and western European firms participation in exploiting oil and natural gas fields in Russia.

For example, the prospects for Nabucco are clouded. Some observers believe Russia developed South Stream, which would take a similar route as Nabucco, as an effort to undermine the rival pipeline. South Stream has the support of Italy, Bulgaria, Serbia, and Hungary. The Austrian state-controlled energy firm OMV has agreed to sell a 50% stake in its Baumgarten gas storage and distribution center to Gazprom. The move could allow Moscow to block Nabucco, as Baumgarten is the planned terminus of the project.[35] In March 2009, in what some observers viewed as another possible effort to block Nabucco, the Russian firm Surgutneftegas acquired from OMV a large stake in the Hungarian energy giant MOL, a key player in the Nabucco project. The Hungarian government has supported both Nabucco and South Stream.

It is unclear whether sufficient natural gas supplies exist for Nabucco. The United States is strongly opposed to tapping Iran's energy resources, due to Iran's support of terrorism, its nuclear ambitions, its policy in Iraq, and other factors. Political uncertainty in Iraq makes it uncertain whether supplies can be drawn from there in the near future. For the present, therefore, the main US hopes lie with Central Asia and Azerbaijan.

Russia and Iran have placed an obstacle in the way of US-supported efforts. They have asserted that no country bordering the Caspian Sea can legally undertake projects such as building a pipeline on the seabed or drilling for oil and gas there without the consent of all Caspian littoral states. Although the other littoral states reject the Russo-Iranian view, this position could make potential investors leery of investing in such a project. Shipments via tanker across the Caspian to Baku can be increased, but would be more expensive. Long-standing tensions between Azerbaijan and Turkmenistan over their Caspian Sea border could also hinder US plans.

On the other hand, Russia's own pipeline routes face supply questions as well. Current Russian gas fields in Siberia are in decline, and the development of new ones, such Shtokman in the Barents Sea and the Yamal Peninsula, have been hit by delays. In the past, Russia has been able to make up for any supply shortfalls by delivering supplies from Central Asian countries. However, Russia's grip on Central Asian supplies has been called into question. A dispute erupted between Russia and Turkmenistan over responsibility for an explosion in a gas pipeline in Turkmenistan in April 2009, while led to a halt in Russian gas imports from Turkmenistan. Supplies resumed in 2010, albeit at a reduced level.

Turkmenistan is moving to diversify its supply routes by developing pipelines toward China and the rest of Asia. A pipeline from Turkmenistan to China opened in 2009, delivering 30 bcm of gas per year. Turkmenistan has also expanded its gas pipeline capacity to Iran, which is expected to reach 20 bcm eventually. Turkmen leaders have also expressed interest in providing gas via Nabucco, if it is built, but no commitments have been made.

Azerbaijan is the most eager advocate in the region of the US-supported pipeline plans, but has the disadvantage that it has only modest amounts of natural gas to export at present. The United States hopes that this will change by 2015 or 2016 as Azerbaijan develops its Shah Deniz gas fields.[36] However, Azerbaijan has shown impatience with delays in the Nabucco project. In March 2009, Russia and Azerbaijan signed a memorandum of understanding to sell Azerbaijani gas to Russia. The decision could deal a serious blow to Nabucco, depending on the volumes of gas that are exported to Russia in the future. The signing of the intergovernmental agreement on Nabucco was a step forward, but prospects for the pipeline continue to be clouded by the failure of Azerbaijan and Turkey to come to terms over pricing of gas supplies.

Other Policy Issues

The Europeans, supported by the United States, may be able to take other steps to diversify their energy supplies. Oil and natural gas pipelines in Europe run in a mainly east-west direction. More north-south interconnections within Europe could help to buffer any shortfalls in a particular region, such as occurred during the gas crises of 2006 and 2009. Larger storage facilities could also be helpful in this regard. European leaders have endorsed European Commission proposals to enhance interconnections within Europe and increase storage. In 2010, the European Union agreed to allocate 1.39 billion Euro ($1.9 billion) for projects to improve the interconnectivity of European natural gas pipelines. Included in the amount is 200 million Euro for Nabucco.

Liquefied natural gas (LNG) delivered to terminals throughout Europe may be an economical substitute for natural gas from Gazprom pipelines. Interconnections and LNG could help create an integrated gas market in Europe. This would make it easier for countries to diversify their supplies and find the best price, rather than have to rely on long-term "take or pay" contracts signed with Gazprom. These contracts require consumers to buy at least a minimum amount of gas, even if they do not need it. Moreover, although the price for the gas is determined by a formula that includes the market price for oil and other fuels, these contracts can result in prices that are higher than in the growing "spot" market for gas.

Already, LNG (largely from North Africa) makes up 15% of Europe's gas imports, and is particularly important for some western European countries. For example, 65% of Spain's gas imports are LNG..[37] Poland intends to build an LNG terminal on the Baltic Sea in order to diversity its supplies. It hopes to have the terminal in operation by 2014. The new terminal will be able to serve countries in central Europe and the Balkans as they pursue efforts to interconnect their pipelines. Another option for these countries is a planned LNG terminal on the Adriatic Sea in Croatia.

Advances in tapping "unconventional gas"(that is, gas from sources such as shale deposits or coalbeds that once had been viewed as too difficult technically or too expensive) could result in increased LNG production from a wider variety of suppliers, perhaps resulting in a further erosion of Russia's leverage. The United States is a leading producer of unconventional gas. US and European officials have stressed the need for countries of the region to improve energy conservation and develop alternative fuels, although they acknowledge that these efforts will take time.

In the long run, Russia's statist, manipulative approach to energy policy may eventually be moderated by its own needs. Some observers believe that Russia will need Western investment and expertise to fully exploit new oil and natural gas fields in remote areas as current ones decline. This may provide an opening for the United States and other countries to persuade Russia to liberalize its energy sector. On the other hand, the current statist system has provided Russia's leaders with immense personal wealth and power, and afforded them the satisfaction of overseeing Russia's renewed international strength. They may not want to change this system, even if it might be in Russia's long-term interest to do so.

In addition to diplomacy, the United States has other tools to deal with the energy dependency question. The United States has funded feasibility studies for some pipeline routes through the Trade and Development Administration (TDA). For example, in August 2007, the TDA provided $1.7 million for feasibility studies on building both an oil and a gas pipeline across the Caspian Sea to link to the BTC pipeline and the South Caucasus gas pipeline.[38] The Export-Import Bank has also provided funds for pipeline projects. The United States and EU are working with Ukraine to develop an energy efficiency action plan for that country. The United States and EU have also urged Ukraine to improve its investment climate to encourage foreign investment in tapping Ukraine's own considerable natural gas and oil resources. The United States provides small amounts of aid to the countries of the region to help build their energy security.

Congressional Response

Members of Congress have expressed concern about the impact of Russian energy dependency on the countries of central and eastern Europe. Members have sharply criticized Russian policy and called on the European Union to work with the United States in helping these countries diversify their energy supplies. Senator Richard Lugar has called for a greater NATO role in energy security issues, including providing emergency energy assistance to member states facing a sudden energy cutoff.[39] Bush Administration officials said NATO could play a greater role in the security of pipelines and other energy infrastructure, but that broader energy issues are best dealt with in other venues.

In the 111th Congress, the Senate Foreign Relations Committee has held hearings on policy toward Russia in March 2009 and energy security in May

2009. In both hearings, Senator John Kerry, Chairman of the committee, Senator Lugar, and witnesses before the committee expressed concern about Russia's use of energy supplies for political purposes. In July 2009, the Senate Foreign Relations Committee held a hearing on energy security during which Senator Kerry and Senator Lugar hailed the signing of the intergovernmental agreement on Nabucco, at which Senator Lugar was present. Nabucco was also discussed at an August 2009 Senate Foreign Relations hearing on Georgia.

In February 2009, in a hearing of the House Foreign Affairs Committee on US-Russian relations, Chairman Howard Berman and several witnesses criticized Russia's political use of energy in Europe. In hearings of the Europe Subcommittee of the House Foreign Affairs Committee in June and July 2009, Members expressed concerns about Russia's energy policy in Europe with Philip Gordon, Assistant Secretary of State for Europe and Eurasia, who repeated Administration calls for European energy diversification. In a March 2010 hearing on Trans-Atlantic security, Chairman Berman and several witnesses referred to energy security as an issue in the relations between Europe, the United States, and Russia.

Congress has also passed resolutions that refer to concerns about Russian energy policy. S.Res. 530, in a list of criticisms of Russian policies on the eve of the St. Petersburg G-8 summit in July 2006, expressed disapproval of Russian energy policy toward Ukraine, Georgia, Moldova, and other countries. H.Res. 500, passed in July 2007, charged that Russia and other countries in creating a gas OPEC, and criticized Russia's use of its gas supplies as a political tool against Georgia, Ukraine, Belarus, and other countries. In July 2008, the Senate passed S.Res. 612, which called on the United States and other countries to build a constructive relationship with Moscow at the 2008 G-8 summit, but criticized some Russian policies. It said that "the conduct of Russian trade and energy policy has created a widespread perception that the Government of the Russian Federation is using oil and gas exports and economic policy as a means of political pressure on countries that seek closer ties with the United States and Euro-Atlantic partners."

In the 111th Congress, Representative Ros-Lehtinen introduced H.Con.Res. 61, which calls on Russia's G8 membership to be conditioned on its compliance with its international obligations and commitment to democratic standards. The resolution says that the Russian government has apparently used its energy resources as a political tool against neighboring states, harassed domestic and foreign energy companies to gain control over these industries; and tried to gain control over energy infrastructure in Europe in order to dominate European energy markets.

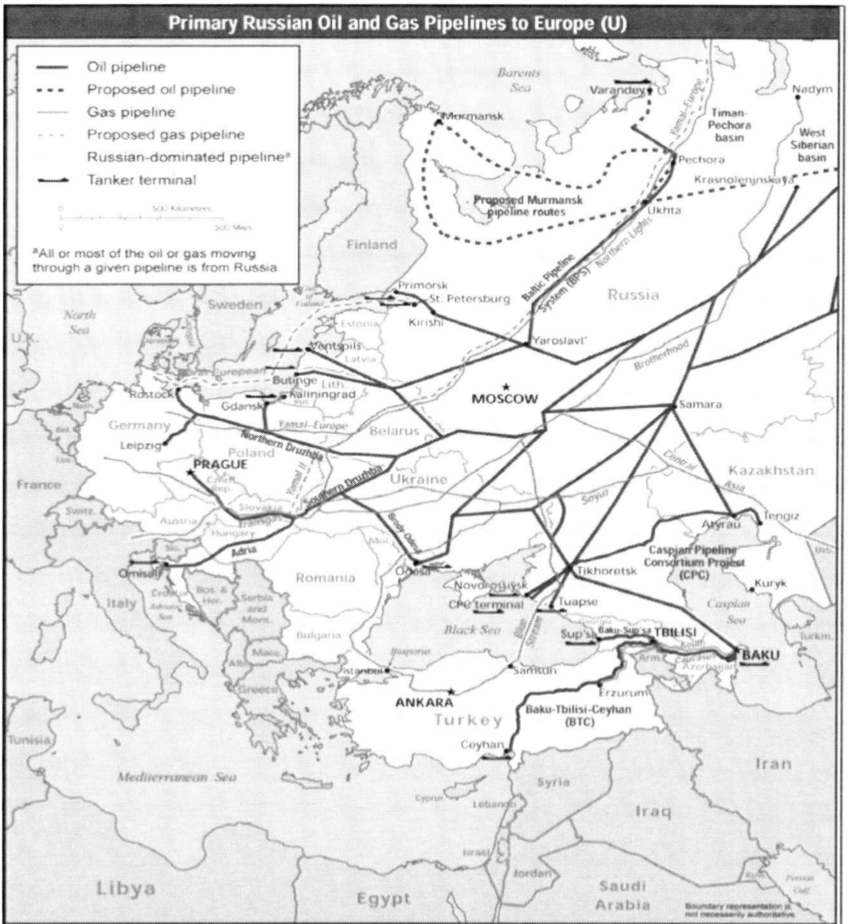

Source: US Department of Energy, Energy Information Administration.
Country Brief: Russia, April 2007, form the EIA website http://www.eia.doe.gov

Figure 1. Pipeline Map.

End Notes

[1] Data is from the Eurogas website at http://www.eurogas.org/figures_statistics.aspx .

[2] US Department of Energy, Energy Information Administration, Country Brief: Russia, May 2008, from the EIA website http://www.eia.doe.gov.

[3] EU Energy Portal website http://ec.europa.eu/energy.

[4] This report focuses on Russian oil and natural gas industries, due to their central importance to the countries concerned. Other energy issues, such as Russia's nuclear power industry, as well as electricity and coal exports, are not dealt with in this report.

[5] EIA Country Brief: Russia, May 2008, from the EIA website http://www.eia.doe.gov.

[6] "France Wants Stake in Russia Gas Giant," Reuters news agency, October 10, 2007.

[7] "Peter Finn, "Russian Giant Expands Control of Oil," Washington Post, September 29, 2005, D06.

[8] For background on Russia's oil and natural gas industries, see CRS Report RL33212, Russian Oil and Gas Challenges, by Robert Pirog.

[9] For background on EU energy policy, see CRS Report RL33636, The European Union's Energy Security Challenges, by Paul Belkin.

[10] Vladimir Socor, "South Stream; Gazprom's New Mega-project," Jamestown Foundation Eurasia Daily Monitor, June 25, 2007.

[11] Keith C. Smith, "Russian Energy Pressure Fails to Unite Europe," CSIS Euro-Focus, January 24, 2007.

[12] Testimony of Zeyno Baran before a House Foreign Affairs Committee hearing on "Central and Eastern Europe: Assessing the Transition," June 25, 2007.

[13] The text of the National Security Strategy can be found at the website of the Russian National Security Council at http://www.scrf.gov.ru/documents/99.html.

[14] US Department of Energy, Energy Information Administration, Country Brief: Ukraine, August 2007; and Country Brief: Russia, May 2008, both from the EIA website http://www.eia.doe.gov.

[15] International Energy Agency, Ukraine Energy Policy Review 2006; from the IEA website at http://www.iea.org/ textbase/nppdf/free/2006/ukraine2006.pdf

[16] Roman Kupchinsky, "Russia/Ukraine: Pipeline Conflict Resurfaces," Radio-Free Europe Radio Liberty Newsline, June 28, 2007.

[17] Glenn R. Simpson and David Crawford, "Supplier of Russian Gas Draws Investigation," Wall Street Journal, April 21, 2006, 1. For background on the gas crisis, see CRS Report RS22378, Russia's Cutoff of Natural Gas to Ukraine: Context and Implications, by Jim Nichol, Steven Woehrel, and Bernard A. Gelb.

[18] The Times of London, September 15, 2007, 4.

[19] International Crisis Group, "Moldova's Uncertain Future," August 17, 2006, from the ICG website http://www.crisisgroup.org.

[20] For more on Georgia, see CRS Report RL33453, Armenia, Azerbaijan, and Georgia: Political Developments and Implications for US Interests, CRS Report RL30679, Armenia, Azerbaijan, and Georgia: Security Issues and Implications for US Interests, and CRS Report 97-727, Georgia [Republic]: Recent Developments and US Interests, all by Jim Nichol.

[21] US Department of Energy, Energy Information Administration, Baltic Sea Regional Fact Sheet, July 2006, from the EIA website http://www.eia.doe.gov.

[22] Eurasia Daily Monitor, Volume 3, Issue 161, August 18, 2006.

[23] Keith C. Smith, "Russian Energy Pressure Fails to Unite Europe," CSIS Euro-Focus, January 24, 2007 from the CSIS website http://www.csis.org/europe.

[24] Russian authorities cited track repairs and a shortage of wagons. RFE/RL, Newsline, May 3, 2007.

[25] Economist Intelligence Unit, Country Report: Belarus, February 2010.

[26] For more on Armenia, see CRS Report RL33453, Armenia, Azerbaijan, and Georgia: Political Developments and Implications for US Interests, and CRS Report RL30679, Armenia,

Azerbaijan, and Georgia: Security Issues and Implications for US Interests, both by Jim Nichol.

[27] The State Department. Statement, January 1, 2006; Daily Press Briefing, January 3, 2006; Secretary Condoleezza Rice, Remarks at the State Department Correspondents Association's Breakfast, January 5, 2006.

[28] "Vice President's Remarks at the Vilnius Conference," May 4, 2006, from the White House website http://www.whitehouse.gov.

[29] Testimony before the Commission on Security and Cooperation in Europe of Gregory Manuel, Special Advisor to the Secretary of State and International Energy Coordinator, Matthew Bryza, Deputy Assistant Secretary, Bureau of European Affairs, and Steven Mann, Principal Deputy Assistant Secretary, Bureau of South and Central Asian Affairs, June 25, 2007.

[30] Secretary of State Condoleezza Rice, "Opening Remarks at the Office of the Historian's Conference on US-Soviet Relations in the Era of Détente, 1969-1976," from the State Department website, http://www.state.gov.

[31] "Statement on Agreement to Restore Gas Flows to Europe," January 22, 2009, http://useu .usmission.gov.

[32] Eurasia Daily Report, October 12, 2007.

[33] Testimony of Gregory Manuel, Matthew Bryza, and Steven Mann in a hearing before the Commission on Security and Cooperation in Europe, June 25, 2007.

[34] Morningstar's testimony before the Senate Foreign Relations Committee hearing "$150 Oil: Instability, Terrorism, and Economic Disruption, July 16, 2009; State Department Foreign Press Center Briefing, June 23, 2009.

[35] Eurasia Daily Monitor, November 19, 2007.

[36] Testimony of Gregory Manuel, Matthew Bryza, and Steven Mann in a hearing before the Commission on Security and Cooperation in Europe, June 25, 2007.

[37] See CRS Report RL33636, The European Union's Energy Security Challenges, by Paul Belkin.

[38] "Promoting Technology in the Oil and Gas Sector," TDA website, http://www.ustda.gov.

[39] "Remarks to the German Marshall Fund Conference," Congressional Record, December 7, 2006, S11483-S11485. For a discussion of NATO's role in Western energy security, see CRS Report RS22409, NATO and Energy Security, by Paul Gallis, updated regularly.

Index

C

D

E

F

G

H

N

O

P

R

S

T

U

V

W

Y